The R.A.M.S. Library of Alchemy
Volume 18

The Last Will
and
Testament
of
Basilius Valentinus

Monk of the Order of St. Bennet

Edited by Philip N. Wheeler

RAMS
Publishing Company

R.A.M.S. Publishing Company
1112 Alston Court
Waynesboro VA 22980

The R.A.M.S. Library of Alchemy, Volume 18:
The Last Will and Testament by Basilius Valentinus
Copyright © 2017 Althea Productions LLC
All rights reserved.

First Edition 2017

ISBN-13: 978-1543151619
ISBN-10: 1543151612

Image Processing by Philip N. Wheeler

Printed in the United States of America

Dedicated to Hans W. Nintzel,
American Alchemist
and
Founder of the
Restorers of Alchemical Manuscripts Society
(R.A.M.S.)

Contents

THE LAST

VVILL

AND

TESTAMENT

OF

Basil Valentine,

Monke of the Order of
St. BENNET.

Which being alone,
He hid under a Table of Marble, behind the
High-Altar of the Cathedral Church, in the Imperial City of *Erford* : leaving it there to be
found by him, whom Gods Providence
should make worthy of it.

To which is added
TWO TREATISES

The First declaring his Manual Operations.
The Second shewing things Natural and Superna-
tural.
Never before Published in *English*.

LONDON,

Printed by *S. G.* and *B. G.* for *Edward Brewster*,
and are to be sold at the sign of the *Crane*
in St. *Pauls Church-yard*, 1671.

The Last Will
and
Testament

of

Basil Valentine,

Monk of the Order of St. Bennet

Which being alone,
He hid under a Table of Marble, behind the
High-Altar of the Cathedral Church, in the
Imperial City of Erford: leaving it there to be
Found by him, whom God's Providence
Should make worthy if it.

To which is added
Two Treatises
The First declaring his Manual Operations.
The Second showing Things Natural and Supernatural.

London,
Printed by S.G. and B.G. for Edward Brewster,
And are to be sold at the sign of the Crane
In St. Paul's Church-yard, 1671.

Disclaimer

Liability: The publisher does not warrant or assume any legal liability or responsibility for the accuracy, completeness, or usefulness of any information, apparatus, product, or process disclosed. The publisher makes no representation as to the accuracy or completeness of the contents of this book and specifically disclaims any implied warranty of merchantability or fitness for any purpose. No warranty may be created or extended by written sales materials or sales representatives. You should obtain professional consultation where appropriate. The publisher shall not be liable for any loss of profit or other commercial or personal damages, including but not limited to special, incidental, consequential, or other damages.

Introduction
Philip N. Wheeler

I started editing this edition of Basil Valentine's **Last Will and Testament** in the usual way: start with the R.A.M.S. edition and correct all obvious errors, especially those introduced by the optical character recognition process. However, the text perplexed me in so many places, even in the first chapter, that I decided the best course of action would be a complete word-by-word comparison with the original edition printed in 1671, starting with Book 1. This was very time consuming, but I am confident that the result will prove my efforts to be worthwhile. To be clear, this is not a word-for-word transcription of the 1671 edition: it is the corrected version of the original R.A.M.S. edition.

Basil Valentine's **Last Will and Testament** contains countless explanations and clarifications of his earlier writings. From the preface, "[This is] the most principal, excellent, and plain of all his works; and I may justly say, of all the Books that are extant on this subject."

Of particular interest are his explanations of **The Twelve Keys of Basil Valentine**,[1] which have long perplexed researchers into the art of Alchemy. In this section, Basil says, "The things I do write of, I know experimentally to be true."

The text contains details of his actual experiments, both successes and failures, including the preparation of the Stone of the Philosophers. There is a step by step explanation of the process for the preparation of the Aurum Potable, in which Basil writes in Part 5, "My intention is to make a perfect relation of aurum potable, for the benefit of good and understanding men." Another type of aurum potable is also presented, and a half-aurum potable, with two methods for its preparation, plus the preparation of potable silver.

The Manuals, or manual operations, are extensively described after Book 5. One example is the preparation of the sweet Essence of Vitriol, and some of its uses are described: "This Essence of Sulphur, four grains of it being taken in Balm-water, dries up the bad humors of the blood, strengthens and incites Men and Women to Copulation, cleans the womb, hinders the Rising of the Mother, and breeds good Seed for the Procreation of Children."

The Last Will and Testament is an amazingly clear work on Alchemy, and I recommend it highly.

---

[1] Volume 1, The R.A.M.S. Library of Alchemy.

To the Reader

If we reflect on former ages, and consider how few there were of such that employed themselves in the scrutiny of the Secret Arcanums of Nature, and how destitute the Philosophers of those times were of those helps and advantages that conduce to the right understanding of such mysteries which we enjoy. That *Hermes* the Prince of Philosophers seemed to be alone, and the only inventor of this most excellent Art. *Cremerus* the Abbot, and our Country-man, after thirty year's study, and extensive practice and labor, was compelled to seek in Italy for a Master Flamel, the Lutetian did partake of the same hard fortune, together with many others of the Ancients. Our Basilius, confesses that he should never have attained unto the knowledge of this Art, if by God's special Grace, some Books of the Ancient Masters had not come into his hands.

Have we not then cause to bless God that we are in such an Age, and in such a Nation also wherein we may converse with many, both Ancient and Modern Philosophers, such whose works do speak then to be Servants to the most High, Beloved of Him, and acquainted with the most occult and Secret Mysteries and Arcanums of Nature, that having toiled and spent much of their precious time in fruitless labors, to prevent the like in their successors, have discovered, and as it were, chalked out the way (so far as was lawful for them to do) how the ingenious may attain to the True understanding of this noble Art: Such is our Basilius, who from his infancy was dedicated to the Service of God; and did so well improve those Spiritual and Natural Talents he was entrusted withal, that there was not anything he attempted the knowledge of, that was concealed from him.

So charitable and liberal a spirit was he, that he became useful and helpful to his Brethren, not only in curing their Maladies, but instructing them in the knowledge of Nature's Secrets: Also he could not go to his eternal rest in peace, unless he imparted unto posterity the means and ways how he attained unto so great Mysteries. Which Mysteries, and the several Processes and ways of Operation you will find in this Subsequent Treatise, being the most principal, excellent, and plain of all his works; and I may justly say, of all the Books that are extant on this subject.

It is one of the greatest unhappiness's that accompanies Man-kind, that there is such a Babel of Languages, that every Language is not understood in every place, this alone prevents the discovery and growth of many profound Mysteries; how ignorant were our predecessors of the works of Basilius, because they were wrapped up for the most part in the German Language, very little thereof being in Latin, until of late years this part was done into English, as it is presumed by one that understood not the terms of the Art, by the many gross mistakes committed, as the rendering *Vitrum*

5

Antimonii, the Vitriol of Antimony, and many others, so that out of the love that I bear to the author, and to propagating so ingenious and True an Art and Science, I have corrected great parts of the former Translation, and *de novo* translated the rest, that was so imperfectly done that it would not admit to amendment, more is also added unto it, that was not in the former, the meanness and plainness of the style, pleaded for itself, that the subject is best understood in that dress, and that the Author did effect it, being plentifully endowed with the grace of humility, offering things more than words, as appears throughout his writings. Read him diligently and often, for the oftener you read him, the plainer do his works appear, it being usual with the best of Authors in this Art, to discover the Secret Arcanums most plainly, yet so dispersedly, and in so confused a Method, that the envious and unworthy shall never be able to reduce them into their due order, therefore often reiterate the readings of this Book, and compare his Sayings with the sayings of other Philosophers, and if you are an adept in this Sacred Science, you will most easily apprehend the Manual Operation thereof. I need say no more concerning our Author, if you read his works, they will sufficiently testify his worth.

I heartily wish every one of you, if adept thereunto, as much contentment and pleasure in the reading and operating the several works herein mentioned and discovered by our Author as himself had in the writing and experimenting them. And that your labors and experiences may be crowned with equal success, so that you as devoutly and humbly undertake the Enterprise, as he did, and make so good a use of what Divine Providence shall permit you to partake of as our Author did. Which that your number may increase. to the improvement of true Knowledge and Philosophy, to the advancement of the Medicinal Science, for the health of Mankind, and to the multiplying of riches in your hand, or in the hands of such that may be helpful to the poor and needy, is the daily prayer of,

F.W.

Preface of Basil Valentine

The Preface and Entrance
unto
Basilius Valentinus
His Last Will and Testament

Whereas the time has come, that by the good pleasure of Almighty God my Creator, and his dearly beloved Son Jesus Christ, my Laboratory in this corruptible World draws to an end, and I am to approach from this earthly laboratory unto the Heavenly, and am creeping daily and hourly nearer to the end, and I am to close according to the appointed time of my Dear Savior, aiming at the true Manna, or Heavenly food of Eternal bliss, sighing continually after me enjoying of such riches, which have a fullness of everlasting joys; I thought it my duty in conscience, to be reconciled in a Christian way to my fellow members that are alive at the present, and with those, which are to succeed in future ages, in that my Soul may be at rest, patiently waiting for the Lords call, and to become an inhabitant of that Heavenly Paradise, and to be matriculated into the Book of Life, and stand in readiness day and night to look for my lords coming. At the consideration hereof I call to mind my writings, which I set down to paper, as other Ancient Philosophers have done before me, publishing all such Mysteries of Nature, whereby Artists, and such that bear an affection and love unto such Mystical Truths may be benefitted, and the same I lovingly and readily leave to them, as much as the highest Spagyrick, and heavenly physician has granted and revealed unto me: My conscience, further has pressed me, in the pursuance of a Christian love, and performance of my promise, dictated by Nature, to make a larger relation, because it is a thing meet and necessary to set forth the Manuals which are belonging hereunto, and ought to be described circumstantially, to inform the judgement of such men addicted hereunto to the full,

7

namely how Natures bolts, which she several ways thrusted forward, to lock strongly her Secrets, are to be thrusted back, that the doors of worldly Treasures might be unlocked, that the knowledge of Transcendent Mysteries may be attained unto upon serious continued prayers unto the Creator, men's judgement and understanding might the more be egged on.

I am not put upon either by force or indigence, nor by a vainglory, or self-end, to set down any letter, and to leave it to posterity, only a mere consideration of the frailty and of the miserableness of this world, where the children of darkness are almost quite lost in their groping ways, has caused me to do so. I am not able to express how much my mind is perplexed, when I think on the folly of this frail world, and consider the Cymmerian Darkness of its Children, which think themselves of deep understanding, when they have heard some Sopperies of the Universal Chair-men discoursed of, thereby supposing to be much enlightened in their understandings. Where is that high and precious Medicine of the Ancient Philosophers, which lived before any University men talked, or heard of, who received their Revelations upon an earnest pious prayer unto God, joining thereunto their daily searching into natural things, and laying their hands to the work. Whither I say, is fled their painfulness and industry? Surely to no other place, then to the horrid, impure, and sordid Apothecaries-shops, deeply precipitated and buried in the earth, and sunk so deep, that the ringing nor sounding of bells, and beating of mortars are able to rouse it: Thus, their praise and the Glory of the highest is at stand! And who produces any good of it? Thus much am I persuaded, that if my writings shall be produced to light after my death, and the judicious posterity, and Disciples of them do rightly read, meditate, and understand them well, and do out of a Christian intention seek, dive, and work the same only to the glory of God, and the love and charity of their fellow Brethren and Christians, then by means of their actions and works, that depend, from such good and proper fundamentals, will be deserved and laid open to the public view, all those Sopperies and forgeries, which these great Mouthed vainglorious fellows, and high conceited fools, which proclaim themselves to be the sole physicians and Masters of both Medicines, that is of the inward and outward, and arrogate to themselves great dignities and worth in the world, when there is no cause for it. Their intentions are set upon no other foundations, but to live in great respect, to haunt after vain-glory with a fearful conscience, to deprive their fellow Christians of their monies by cheating, all they look for, is to be talked of, and live in reputation, they slick full of diabolical pride and vanity up to their ears, these in the end, in great wo and misery have their souls drowned most lamentably! Wo, Wo, to you children of Satan! Here I intended not to use any prolixity of words, not to bring in any such matters, which are heterogenic to my purpose, at the closing of my book I will be more large in my expressions, as far as the heavenly Prince shall enable me to do, whereas for a corollary to all my writings, I shall annex things with that proviso and entreaty, that all those, which intend to be real, conscionable in their Medicaments, may aid and cure their oppressed fellow Christians, and to search and inquire into such things, which God for such purposes both ordained, and implanted then into Nature.

This present book of mine deserves to be called (A light unto Darkness) for other things, which in my former writings I have discoursed of by way of parables, which style I made use of the rather, because it is proper to Philosophers, are declared in this my last information, where I deal in plain and clear words, describing, and naming the matter openly. Showing the preparations from the beginning, to the middle, and end, demonstrating, and setting before the eyes of men the matter in general and particular, confirming, and justifying the truth thereof, and making a distinction betwixt the ground and no ground in so plain terms, that the very Children may understand, and feel it with their hands. And because this book affords another knowledge, differing from others of my writing, wherein I have not written so obscurely, nor made I use of such subtleties, as the Ancients did, who lived before me. And ended their days happily, therefore does it require another place also to be laid up in, and kept secret from the perverseness of men in the world. I do not desire it should be buried with me, to be a prey, and food for the worms, but it shall be left above ground, and kept secret from wicked men, and my purpose is, that it shall be laid into a secret place, where none shall come near it, but he, for whom God has ordained it, other writings of mine shall sooner see the public light.

But know that whoever you are, into whose hand this my last Testament comes, which contains the Revelation of heavenly and earthly Mysteries, it will happen to you by a divine providence, to whose custody, is my devotion then performed, I committed it, depositing the same into that secret place together with other things, not enforced upon any grounds of Mysteries, or straightness, to leave it there only for him whom God's goodness shall appoint to find it. For it is not good for me to let God's creatures and mystery, which are too abstruse already, and stepped from the light into darkness, by reason of the perverseness of the wicked world to die with me, as wicked men are wont to do with gifts, they are entrusted withal; but even leave a glimpse of truth and of the clear heaven, thinking befitting to discharge my conscience in putting forth the talent to usury, let the will of the Lord be done in him, whom he deems to be worthy of it, to whose care and diligence I commit it from henceforth and forever. For I, a Cloister-man and an unwearied Servant of the Divine infinite Trinity confess and acknowledge, that I should never have come so far in my knowledge of these endless mysteries, in the Analogy of natural things, in the melioration, changing their qualities, for a sure and strong upbuilding of this profound knowledge of the True Medicine, according to this Ordinance (whereby I am ready to do good to all and every one, which desires my help herein) which as I have done hitherto, so my desire is that God would enable me to do the same to the last of my breathing.

I say I should never have attained unto it, if by Gods special goodness, grace, and mercy, several books had not come into my hands, written of ancient Masters: Which departed this life a long time before me, causing great joy to me, stirring up in me a return of hearty thanks to God, who so graciously was pleased in this providence to bestow them on me in the Cloister before any of my fellows.

I do speak it without any vain-glory, I have done so much good thereby to my fellow Christians, as ever was possible for me to do, which next to God, returned thanks for it, even to the end of my days.

Now, I can be steward no longer, I have done according as my abilities would give leave: Let others also be industrious, and not defective in their stewardship, I return mine into the faithful hands of God Almighty, and deliver up to his Divine power and glory, instead of the Keys for the house, all the allotted mysteries set down here in my writings, leaving them to his disposing, to bestow them after my death unto him, whom his Divine Will, shall chose thereunto, to be for the knowledge of his praiseworthy name, the good and help of such, that stand in want of necessaries and health, for the avoiding and shunning of all worldly pomp, pride, wantonness, luxury, rashness, pleasure, covetousness, and spitefulness, of contempt of others.

O Lord God Almighty, merciful Gracious Father of thine only begotten Son Jesus Christ, who are only the lord of Sabaoth, the principle of all things that are made by thy Word, and definite end of all Creatures above and below; I poor miserable Man, and Earth-born, return thanks with my babbling tongue from the inner most Center of my heart, who has been pleased to enlighten me with the great light of thy heavenly and earthly wisdom, and the greatest mysteries of the created secrecies and treasures of this World, together with thy Divine saving word; by which I learn to know thine Almighty power and wonders. To thee belongs eternal praise, honor, and glory, from eternity unto eternity, that you have bestowed on me health and livelihood, strength, and ability, to be helpful to my fellow Christians in their necessities and infected infirmities with those mystical healing medicines, together wi.th such spiritual comforts, to raise the drooping spirits.

Lord, to thee alone belongs power, might, and glory, to thee is the praise, honor, and gratefulness, for all the mercies and graces you have bestowed on me, and has preserved me therein till to this my great age, and lowest weakness. Of you God of all graces, and father of all comforts, be not angry with me, that I deliver up to thee, my eternal creator, the keys of my stewardship, wrapped up in this Pergamene, according to the duty of my calling and conscience calls for; with these you suffer me to keep house the most of any time, till now, you have called and foreseen me to be your servant and steward, and has graciously afforded, that I should enjoy the noble sweet fruits, which were gathered in your Almonary to my last instant end, which now, O lord, lie in your power. I beseech you for the dear merits of Jesus Christ, come now, when you please, enclose my heart; receive my Soul into your heavenly Throne of grace, let her be recommended unto you graciously, O you faithful God, who has redeemed her on the holy Cross with the most precious Tincture of the true blood of your holy body; then is my life well ended on this earth, grant to the body a quiet rest, till at the last day, body and soul joined again, and are of a heavenly composition; for now my only desire is to be dissolved, and to be with my Lord Christ, the which you Almighty, Holy, and Heavenly Trinity grant to me, and all good Christian believers. AMEN.

That I may come to the work intended., and make a beginning of the work in hand, if God be pleased to let this book come to your hands, before all things incessantly you return hearty, and unfeigned thanks for it; in the next place I beseech him to bestow on you also grace and blessing, a healthful body, successfully to accomplish all the points laid down here, which tend to the welfare and benefit of your Neighbor, and to prepare them according to the Manuals, which to that end I set down, and prescribe them, that you may happily and successfully begin the work; that the middle and end be correspondent thereunto. Then be not flap-tongued, and resolve absolutely in your heart, not entrust with these Mysteries any malicious, ungrateful, and false man, much less should you make them partakers thereof; for if Almighty God, would bestow it on others, he could do it immediately, and grant the same by other means and ways without you. Therefore, look to it, tempt not the Lord thy God, for he will not be mocked. Be silent and reserving; be meditating on Gods punishment, which no man is able to out-run; keep a good intention; let not your greedy mind run on, how you may get abundance of Gold and Silver riches, and vanity, but before all things, which are written herein, let that be your primary aim; how you may appear helpful in word and deed to promote the health of your Neighbor Christian: Then have you given, and brought an offering of thanks, and God will bestow more upon you, and with such Revelations; will come to you; more than ever you would have believed.

Therefore, instead of a perfect Physick book, I have annexed at the end those precious Medicines; with which even to this hour, I have cured successfully many difficult symptoms, and diseases, held by others to be incurable; which I recommend unto you with the rest of the things contained in these writings, to your conscience to be wary, and not to abuse any of them, as much as your life and soul is near and dear unto you.

If you receive this faithful admonition and warning, which I give unto you here and there at the beginning, middle, and end, and in other places also, and do accordingly, then you cannot be too thankful for these things which through God's permission shall be intimated and made known to you out of this book.

But in case you will strive against my faithful warning and fatherly admonition, these mysteries, which were hid from the learned, and then suffered to come into strange hands! Look to yourself, blame not me, think not that there is any reconciliation for you, being cut off from all these, which live and die in Christ.

Thus, I let it rest, committing execution to the Highest, which dwells in Heaven, who punishes and avenges all manner of sin, vices, iniquities, and covenant-breakings. Whereas I thought it necessary thus to describe this my Declaration before my other Writings, and to prove the same with examples, that everyone, whom God shall judge to be worthy of; may conceive, understand, and fathom the true beginning, the true middle, and the true end of all created things. Thereupon I purposed to make a beginning of it with a necessary relation of the Original, beginning, and existence of Minerals and Metals, from which arises the most noble and precious Medicine, whereby is procured a healthful long life, and abundance of riches are obtained: Namely from whence Minerals and Metals have their original, how they be brought to light, that searchers into Natural things may know the whole Nature in her circumference, before ever they lay hand to the work, and acquaint themselves well therewith; then still the one will stream and run forth the other; one art will produce the other; at last all what is fought for, will be overtaken in joy, and that which has been longed for, will happily be enjoyed.

This my book, I divide into five parts, each of them is subdivided into certain Chapters and Parts. As for the style used here, it ought not to be expected to be any other than is meet for a Miner, after the condition of Mine-works; waving all Rhetoric and Poetical manner of eloquent expressions.

I.

In the first part of my intended work I will describe chiefly the matter, nature, and properties of Mine-works, in which grow Minerals and Metals, of the First Sperm, nativity, quality, and property, as also of the exhalation and inhalations.

II.

The second part shall be a kind of recapitulation of the first part, where shall be contained also a relation of the condition and occasion of Mines, Ores, Stones, passages and Clefts, with their coherent liquors, powers and operations, as of Gold, Silver, Copper, Iron, Tin, Lead, Mercury and the rest of Minerals.

III.

In the third part is declared in manifest and Literal expressions without any defect and obscurity, the Universal of this whole world, how all philosophers before me, and after me, have made that most ancient great stone, whereby health and riches were obtained, of the possibility, how, and of what it may be done, together with a perfect Declaration of my XII Keys, with the names of our matter.

IV.

In the fourth part I describe all the particulars of Metals, which the one is endued with, before the other, out of which may be had perfect health, and an advantage unto the getting of great riches, with all the Manuals in general and particular, belonging thereunto; this fourth part I have entitled, the Manuals of Basilius Valentinus, wherein is showed how all Metals and Minerals, fitting thereunto, may be brought to their highest preparation.

V.

In the fifth part I have annexed the transcendent most dear wonderful Medicine of all Metals and Minerals, and of other things, which God Almighty had ordained, and graciously granted for men, who in the valley of misery are subject to sickness and poverty, that they may have a remedy to help themselves against both.

God the Father of mercy and salvation, who lives from eternity to eternity, being above all the Creatures, grant grace and blessing to this my purpose, that I may write so; that everyone may understand, and Gods infinite mercy, that it together with his gracious goodness, redemption my seem known, acknowledged, and continually meditated upon, and everyone may call on the Great Creator day and night, granting to them fervent hearts so to direct all their thoughts, that they may make no otherwise of this noble creature of God and transcendent great mystery of Nature, together with the Anatomy thereof, but only to the great honor of God, and the good of all good Children. The same grant this Father, Son and Holy Ghost in his mercy, AMEN.

Two Treatises Described

OF THE MOST EMINENT AND INCOMPARABLE PHILOSOPHER BASIL VALENTINE FRIAR OF THE ORDER OF THE BENEDICTS.

The First

Whereof declares his Manual Operations, how he has made and prepared his Secret Medicines; the Stone Ignis out of Antimony, and last of all the Philosophers Stone.

The Second

Discovering things Natural and Supernatural, as also the First Tincture, Root, and Spirit of Metals and Minerals, how they are conceived, ripened, brought forth, and augmented.

A Table of Chymicall & Philosophicall Characters wth their significations as they are usually found in Chymicall Authors both printed & manuscript.

Saturne Lead		Balneu Mariæ		Mensis	
		Balneu Vaporis		Mercur: Sopimat	
		Bene		Merc: Sativa	
		Borax		Mercu Sublimat	
Jupiter Tinne		Calcinare		Nota bene	NB
		Calx		Nox	
		Calx vive			
		Calx ovorum		Oleum	
Mars Iron		Caput mortuu			
		Cementare		Præcipitare	
		Cera		Pulvis	
Sol Gould		Christallum		Pulvis Lateru	
		Cinis		Putrificare	
		Cineres clavellati		Putrificare	
		Cinabar			
		Coagulare		Quinta Essentia	QE
Venus Copper		Cohobatio			
		Crocus Martis		Realgar	
		Crocus Veneris		Regulus	
		Æs vstum		Retorta	
Mercury Quicksilver		Crucibulum		Sal comune	
		Cucurbitam		Sal Alkali	
				Sal Armoniac	
				Sal Gemma	
		Dies		Sal petra	
Luna Silver		Digerere		Sapo	
		Dissolvere		Spiritus	
		Distillare		Spiritus Vini	
				Stratu super strat	
		Effundere		Solvere	
Acetum				Sublimare	
Aceti distillat		Finus Equinus		Sulphur	
Æs				Sulphur vive	
Aer		Flamma		Sulphur Philosophoru	
Alembicus		Flegma		Sulphur nigru	
Alumen		Fluere			
Amalgama					
Aurus		Gumma		Tartar	
Antimonium					
Aqua		Hora		Calx tartari	
Aqua Fortis				Sal tartari	
Aqua Regis				Talcum	
Aqua Vitæ		Ignis		Terra	
Arena		Ignis rota		Tigillum	
Arsenicam				Tutia	
		Lapis calaminare			
Atramentu		Lapis		Vitriolum	
Aqua chalou		Lutare		Vitrum	
Aqua zymou		Lutum Sapientiæ		Viri et ceris	
		Magnes		Vrina	
Libra		Scrupulus			
Librae		Dram		Johannes Wolfe	
		Ancia			
		Magnes			
		Marchsita			
		Materia			
		Matrimoniu			

The First Part of Basilius Valentinus

Chapter 1: The Metal Ferch

Of the Aetherial Liquor of Metals, or of the Metal Ferch.

God has created things underground, as well as the things above ground: By the things underground, I understand Metals, Minerals, in which there is implanted also a fertility to their seed, without which the seed could neither grow nor increase. See which is barren, has not that fertility: by which it is collected, that there is some distinction between seed and fertility. If we will enquire narrowly what fertility is, the best and surest way is, to consider life and death of creatures, how they hold together; for death is barren, but a living life is fertile, because it stirs and moves.

It is seen by all the works that are undertaken about metals, that there is nothing so volatile as metals are, and so nothing stirs and moves more subtly than it; but this stirring and moving I will call here the *Ferch* of metals, by reason of its continual proceeding, and incessant moving; and because the same is not visible in metals, and does it in a twofold way, therefore I will let the old word stand, and call its stirring a *Lubricum*, and its *Ferch* a *Volatile*; for with the virtue and power of both these, it performs all that, what it needs for the perfection, purity, and *fixation*[2] of its work.

Seeing *Ferch* is a perpetual living and forthgoing thing, one might admire and say, of what condition is metal then, which we behold with our eyes, and feel with our hands; which being thus hard and coagulated, whether the same be alive or dead, and whether the life or *Ferch* in metals may be destroyed (which is impossible), what is the condition of it, or how comes it to pass? I answer, that a metal may be alive when it rests, as well as when it grows, or stirs; and here a distinction must be made again between the death of metals, and their rest and quietness. For death touches only the bodies, when they perish; but the life itself, or *Ferch* cannot perish or cease; therefore, if a metalline body be extant, then is it at hand visibly two manner of ways. The one is *in liquido*, and is discerned in its moving to and fro, and if it be forced by a strange dangerous heat; then it turns to a volatility, and flies away. The other way is,

---

[2] The original said "*fixation*" but this makes no sense to me; probably a typesetting error. The R.A.M.S. edition has "Sization": clearly an error. -pnw

when it is at hand *in coagulato*, wherein it rests so long, till it be reduced into its *liquidum*, and that is done in a twofold way, and lasts so long as the body lasts, but as soon as the body is destroyed or gone, and is come or entered into a more, either noble or ignoble body, then its *Ferch* or life is gone also; therefore if you will reserve and keep a body, then take good notice of its *Ferch* of life; for if you once stir it, and hunt it indiscreetly, you do it with the loss or diminution of the body, wherein it is, for that life never goes away empty, but still carries along one life or *Ferch* after the other, carrying it away so long, that at the last it leaves none. But what the condition is of the moving and quietness of that life, and how Nature brings it to rest, must be exactly considered. For an accurate knowledge demonstrates, that there is a difference between the life of the seed and of the body; for deal with the seed which way you will, you cannot bring it to a *volatility*, because it is against its kind; and so, the body also is of the same condition; but the *Ferch* alone may be brought to it. For if you provide food for the *Ferch*, then you strengthen its whole work, even as a mother does her child, which she feeds and cherishes well, and brings the same the better to its rest; so is it also with the *Ferch*. Therefore, all such which gaze and view only the seed and body, and know not the fundamentals about the *Ferch*, lose the body, because they observe not Natures progress and proceedings, putting the cart before the horse, or the foremost they put hindmost. This rest and sleep of the *Ferch* serves for that use, because it preserves the body from destruction, or consumption being once come into its perfection. For as long as it is awake so long it consumes, but when it is at rest, then it stands close in a lastingness, and when it has nothing to feed upon, then it corrodes and seizes on its own body, consuming it quite, at last it stirs and moves to another place. Hence is it that treasures, or pagement[3], which are buried, awaken at last, consume their own bodies, reducing them to dust, so that nothing of them remains but either a mere stone or flux, as in many places is to be seen.

---

[3] Verified in the 1671 edition. -pnw

Chapter 2: Of the Seed of Metals

All those Authors which have written about the metalline seed, agree in that, when they say Sulphur is the masculine seed of metals, and Mercury is the feminine seed; which saying must be taken in its genuine sense, for common Sulphur and common Mercury are not meant thereby. For the visible Mercury of Metals is a body itself out of bodies, and so cannot be a seed: and being cold, its coldness *per se* cannot be a seed; and the Sulphur of metals being a food, how can it be a seed? Yea, a seed consumes Sulphur, how can one seed destroy the other. If so, what body should it produce? It is therefore an error, if that should be taken in the common sense: if the Mercury of bodies is in a work, and has taken food, then all the sex Mercuries protrude one body, as the one of the sex is in predominance, so the body rises.

Seeing there are seven of these Mercuries, it happens, that when the seed of *Mars* and *Venus* has predominance, they produce a masculine body of *Sol*, but if the seed of *Saturn* and *Jupiter* has predominated, then is produced a feminine body, which is called *Luna*; *Mercury* is an assistant on both sides.

The same happens to other bodies: but these are always and in every and each work together; for they are indivisible, as it is meet also, what manner of body could be produced else? For Nature has perfect bodies, though in themselves must be dissolved again, yet are they perfect for, and in their time. For what manner of seed could that be, if it should be defective in any of its branches?

Therefore, everybody has its perfect seed, hence the transmutation has its ground in the *ascension* and *descension* of metals, which otherwise could not be, if they were not homogeneal in their seed. For if any man says, that silver is not gold, clowns believe that also, because they have not fundamental knowledge of the seed, how it is to go out of one body into the other, or else it wants its fertility, neither can it be naturally without a body, wherein it rests. There belong Seven distinct parts to a uniform body of metals, to bring it by nourishment into a form, *viz*. 1. An earth. 2. A

Stone. 3. An earth-ash. 4. Earthly streams. 5. Glass or subterranean metal. 6. The subterranean tincture. 7. The Subterranean *fuligo* or seed (fume). All these are the materials of the body; and as earth is man's matter, out of which God made him, unto which he must return again: so, all other bodies also at last return to earth, that Miner which is judicious and knowing herein, him I judge to deserve the name of Miner. For there are but few of them which are rightly informed herein, or have any fundamental knowledge of it, though they are daily employed about it: though some might say, they could not but be knowing it their profession, yet it is not so really; if so, what right use can they make of then? They put wrong names upon them, are ignorant of their utility, and this is the reason why they many times run them waste upon heaps, where after some time they turn to goodness, and the longer they lie there, the better they are: this instruction deserves no hatred, but rather a grateful acknowledgement.

Why should Philosophers be believed to know anything? But where is it written, that men should seek and find mercury of the body in a subterranean fume, stone, glass, but in their books? Where are learned artificial finings infuliginations, incinerations, nutritions, but from them? The seed of metals as it is perfect, so is its *Ferch*, or life invisible. Where do those men stay, which will work according to Nature, and know none of these, neither do they know where to get it: yet fall upon Artists, exclaiming upon them to be false, and all such as are employed in their ways: but we see and hear how ignorance runs on. It is impossible to get a body without seed, it was as much as to say, a seed is without fertility. Therefore, peruse it exactly in its dissolution, the reduction of it will afford its body: work cheerfully.

But it is none of the meanest work, as some of the most ancient Philosophers have said, which called it a double work: for thus they say, that metal must first pass through the Melter's hand, afterwards it must come into the hands of the Alchemist, if so be the seed shall be known in the artificial work: they mean or intimate by this saying thus much, that there is a twofold dissolution, the one is, when the expert Melter brings the frangible body *ex naturali conductione* into a malleableness, whereby its impurity is gotten off. Then comes the Alchemist, reduces the body into cinders, calxes, glasses, colors, fumes, subterranean, in which the seed of metals rests, and the *Ferch* or life is found fertile in the body, and is reducible into a Spiritual water or *prima materia*, according as the quality and property of the metal is, and is divided artificially into its natural principles, according to the process of the Chymick Art, of which more in another place shall be spoken, when I shall treat of the Minerals.

Chapter 3: Of the Metalline Nutriment

Although it belongs not to this place, how *mineralia fossilia* are made under ground, however I will give a hint of it, how nature makes them out of subterranean moist liquors and Mine crescencies, which afterward serve to be a food to metals; not such liquors which are decocted above ground; therefore if you should add here above ground, decocted ones to metals, undissolved in their corporeal form, your work would be in vain: and where there are such *mineralia fosselia* there are Mine-works also if not with it, yet are they not far off, as is seen in many Mines. As in *Hungary* are digged the fairest and best sulphur-alloms and Mineral or Mine Victriol. And about *Harcyria* are digged Salt Victriol. About *Goslar*, *Mansfield*, *Zellerfield*, and *Eshland* in *Helvetia* is digged Mine-salt, and at *Hall* there is great store of it, where there is found also very curious Sulphur. But you must note, that these minerals are not used thus grossly, but are prepared first, which is a curious work to bring a mineral thus high by subliming into *flores*, which are half metalline, especially if made with metal, the metal being reduced into a mineral, from that the flowers are made: thus, you see Natures forwardness, and how she is reducible to her first water, Sulphur and Salt. Many make these flowers without metal, which are not so good though, as the former way. For an *oleum* made of Vitriol or Copper, and is distilled, is more effectual, yea a thousand times more precious in its operation, then that is, which is made of common Vitriol, whom Nature has not yet exalted. Its true, the *Hungarian* Vitriol, in its efficacy and virtue is found wonderful and sufficient enough, because Nature has graduated it to a greater siccity, and brought it to a ripeness more than others were, and is more excellent than the rest. By this preparation, they can make use of the minerals, strengthen and increase their pleasure thereby. If anything is to be made meet for metals, then it must be done out of metals, with metals, and through or by metals, which is the real and only manual whereby may be hit the hardness of the mineral flowers, always take from them, and add nothing to them, this is the Art, which asks great wondering, and deep meditation. Thus you must learn to go to work, for these flowers are found often closely compacted, which Miners very seldom know, especially in *Hungary* and *Wallachia*, they are as fair as ever any red glowing ore may be, they are of a crystalline transparent redness, are good gold and silver according as they are tinged, this is a rare knowledge, an art worthy the best consideration, which is to make glass of a hardness, from thence it is, that the subterranean glasses make up the metal, thereby they come to their own form.

The preparation of these flowers have their great utility in physical ways, if their excrements be taken from them, and their odors: these excrements are the faces of minerals, are naught for metals, stirring up evil sediments, which bring damage unto metals, a twofold evil comes from the mistaking of minerals: for decocted ones are a dangerous poison, and corrosive unto metals, as we see above ground, when *aqua fortis* is made of them, which corrodes, tears, divides, and parts metals, and the other which are fair to look on, sticking unto metals, and their worst poison, for as soon as these approach, they kindle and cause the dangerous sediments, all avails nothing unto them, though they have and keep their form. As an infected man, has still the form and face of a man, though he be infected and infects others also, and in case it turns all to one metal, yet it is but an empty one, and nothing in it. This is a very necessary observation for Miners and Laborators[4], for if they regard it not, they obstruct not only their work, but endanger themselves also; because the metal is not only turned into a volatility, if any feces or excrements be added thereunto, and that also which stays, comes to be unmalleable, and suffers continually diminution, as long as it is under the hammer. Those that work them, have cause to look to it, if they fall on them with any fire, their reward surely is some mine disease, which experimentally is known how their poison does stick and hang on the top of the furnace and in their chests, turning to arsenic and such poisonous fumes and seeds, and do hurt every way, as woeful experience evidences.

---

[4] Probably "Laboratory Workers" -pnw

Chapter 4: Of the Metalline Shop
Officina Metallorum

All natural works have their special convenient places in which they work; where there is any such place or shop, in which some glorious and precious thing is made; and sometimes though the instrument be very horrid and monstrous, and its matter unknown, yet they are extant in that office.

First, touching, the glory and praise of this officine, it is likened to a Church, in which the seed and the *Ferch* are married to the body, therein they eat, rest, and work, thither they carry all fair and pleasant materials under ground wherewith they are clad, and they have another kind of fire, water, air, and earth, for the things that are accomplished and perfected therein, the same can hardly be parted again, no not with the help of the nether air, if so be, that it must be parted asunder, then see and make trial of it on the mercury of metals.

Again, the things made so hard and fixed cannot be parted, as may be seen of gold, how firm and fixed is it in the fire? The cause whereof is the subterranean heat and cold which it imparts to metals, and makes then firm thereby, for it is a stony firmament of the earth, and giving to metals their stony power, it grows hollow and spongy, full of pores, which at last are filled up with metals, even as Bees do fill their hives with honey, and in the end, it parts, and is carried away in the slick, or (*Scobes*). For the Earth-stone is not consumed underground, because it is a sediment, not suffering anything to come in or out, hence is that difference betwixt the Earth-stone and the terrestrial-firmamental-stone, which is one of the mineral-works. Let no man gainsay, that a stone should have together both heat and cold at once, to afford the one now, and then to hide the other, for when it works upon inferior metals, then it hides its coldness, and so it helps every way, this is its tract and instrument, heat and cold of the subterranean fire-stone.

The modern Chemists which are ignorant, not knowing Nature aright, and do not take notice of her ways, use strange instruments, and then they make or cause to be made all manner of vessels, according as every one of them have a fancy to, but in natures ways they know little, she regards not the variety of forms, and instead of these, she takes a fit and lasting instrument, which holds in the work, and every form follows or accompanies the seeds precedency. The folly and ignorance of workmen is aggravated in that because they despise the knowledge of minerals.

The instrument she uses hereunto I should make mention of it here, but wave it at this time, and will do it in another place, where you may seek for, and take notice of it. Those which think themselves to be the wisest do say, that it is a vanity to observe mathematically the stars above, and to order any work after seasonable days and hours, it is something said, but not so well grounded. But this is most certain, that if you work according to common course, otherwise than we do, following only your own fancies, then is your labor in vain. There is a difference to be made between the upper stars, and the metalline stars, which shine and have their influence into bodies. Touching the stars above, they in their light and notion have a singular influence; and the stars below have their influence also upon their metals, thus each heaven has its peculiar course and instrument, where the stars situation may be apprehended. An *opacum corpus stellatum* compact astral body, differs in its condition from a *corpus lucidum*, if you intend to learn here something, then you must be industrious and grudge no pains, it would require a huge volume, if I should describe *particularly* the whole circumference of subterranean Mine-works: it would not suffice to nominate the things only, but must demonstrate also, that all that, which I attribute to them, to be true, I say it would ask a great deal of writing, to dispose the brains of misconceited men to a belief; what should I say of such materials, on whom I could not impose fitting names, though I know them, for who is that man which has done learning in our School? Here I must needs speak as belonging properly to this place, that no volume in this world can be written, in which could be set down all and every *particular manual*, as Laborators sometimes might ask; therefore, an Artist having given him some hints of things, must endeavor to order his work, and manage the same judiciously, must put his hands to the work, and get knowledge by his own practice. I direct such men in their work to Natures process underground, let him search there, and take an honest Miner along to show him her instruments, and matters (for prating, lying, and ignorance avails here nothing). Every one wishes to get riches, but the means for the getting of them are not respondent: if I were the best limner, and could set forth in colors the form of any instrument, then men would understand it; it would help in this case they would see it, and feel it with their hands and undertake the work, if all were set down. I know what and how much ought to be put in a book, I put things fitting in, and did it faithfully.

Chapter 5: Of Egression and Ingression of Metals

The work of metals evidences a perpetual going in, and coming out, for hereby the Egression is understood not only the Egression of the whole work, that in some place a whole metalline tract comes into decay, when it wants food to be nourished any further, and has devoured all its bodies, but also a partial egression; for still the one seeks the other, and follows at the heel. This we see by the mercury of metals, being poured forth, it is scattered into thousands of little quick corns, all of them return to their body; in the same condition is *volatile lubricum*; and the *Ferch* also goes forth in small bits, at last it joins in a body somewhere, even as Bees meet together: it receives no more than it has need, the overplus swarm to another Mine-officine, which parting and distributing, affords many and several Mine-works, according to the disposition of the officine and nourishment, and according as it is infringed in its work in the egression; the *Ferch* and the seed go on in their *volatility*, and if they had wings, that *Volatile* is so thin, that it can hardly be discerned, yet is it foliated like a heap of atoms; thus subtly it flies away, and the *Ferch*, must still have its *seed*, the seed its *body*, and that its thin *atoms*. My meaning is not, that its egression is from or out of the earth into the air, to fly about there, and then to come into ground again; which is not so, nor can it be, because its natural work is not in the air, unless men bring it forth purposely, then is it of another condition; of that egression I do not speak here, because it is done by day; but this goes through the earth. Which stands in the furnace, not apprehensive or visible to us, and runs through clefts and passages. For if the earth gives way to the ingression and egression, even as the water does to fishes, and the air to birds, as long as metals come to their stone-firmament, which stone-firmament differs from the earth-firmament, when it meets with that, it goes about, looking out for another passage, like as water that flows about a stone, and not through it, yet it stays in its own stone and receives strength of it, and turns there to a body: and as it goes in its egression from one metalline firmament, stone-firmament to another, if thorow eaten soaked[5], be it at what distance it will, attracts the *Lubricum*, even as a bird draws it feet up to its body in its flight; for if they touch anywhere, then they lose somewhat of the body, and the *Lubricum* in its ingression suffers it to come again to a strength of operation; for when both are joined, then the metal increases, and attracts its food in a wonderful way, and nourishes itself, and it is to be admired, that in this ingression, when that *Lubricum* comes more and more to its officine, how it increases and strengthens itself so long, that at last the work is made firm in the officine. This strengthening cannot be learned to be any other, then the metalline mercury does make it, for in the first place it turns it there into a *liquidum*, where afterwards it receives all, does coagulate and congeal, according as the bodies are either *masculine* or *feminine*, at last it is brought to a solid fixed body of *Sol*. This ingression makes that subterranean place noble and fruitful, and is singular, when it has an ascending ore in work, that air is very wholesome, and if the air above with melting be not infected with arsenic fumes, then it affords a saluber[6] air to dwell in.

This is a manuduction into the whole afterwork, how the same ought to be proceeded in, that ore may stand and not awaken, but turn to its stream, and still abide in its bodies company, it is loath to make an egression, if once it made a true ingression, and settled itself to the work, for it rests not in its place, neither does it rest in its whole tract, but works continually, and is well seen, what its fixing or flights is, and where it sets to a fluid body, or earth-salt, which it stirs and rules so long, yea, it pants and moves in it so long, till it gets a liquid body, then turns it to a terrine body, and is still brought on to a further height and hardness: and that is the right coagulating, congealing, liquidating, and fixing of mercury; which if done accordingly, then it affords something.

---

[5] This is how the original text was printed. -pnw
[6] This is how the original text was printed. -pnw

Chapter 6: Of the Dissolution and Reduction of Metals

It is apparent, that natural heat is the cause of the fluidness of metals dissolution, because the seed of metals in itself is very hot, and the fluid matter of metals is hot also, as being oleaginous, and its heat increases, when it comes to its officine, or shop, because that also being hot increases the heat more, hence it is why it is hot in the work, and has need of it, for at first it would bring no more into its body, unless it were soluble and soft, it brings nothing into it, unless it be passed through these three heats, and fixed by them: then examine it, and add another fluid thing to it, which did not pass through the three heats: see whether the metal will receive it, or no? Secondly, they must be dissolved, that they may be cleansed; the condition of liquid things is to produce to the outside things fitting the work it has in hand. This solution is distinct from other artificial dissolutions, where the body is only melted, as by the Melter when he separates the excrements from it; for nature does not melt the earth as men do, but as corn grows above ground, so she leaves corn and husks together; there is a great difference between our melting and the dissolving of Nature; if we could observe that distinction in our dissolutions and meltings, we should not be at so great losses and damages as we are: I must needs mention about *aurum potable*, how men do busy themselves about it, as many heads, as many ways they choose to the making of it. Some take that which is not yet separated from the metal containing yet the cinders of excrements, or worse things. They take corrosive waters, *acetum aquavitae*, and the like: pray, tell me, what does Nature take when she is about the dissolving of a congealed water? She takes none of these things, only makes use of a heat. You must do the like, if you will take a metalline body, which Nature has perfected, and through melting and fining is come to us, if you will dissolve and reduce it to its first matter, then rouse the *Ferch*, thus you may make any metalline body potable, being made pure and superfine, then its excrements are gone, made not with additionals of corrosive things: the fluxing of such matters rather make the metals harder: if a body shall be fixed, we fix it from without, which Nature does not, for she fixes the seed, then the flour sets and turns to such a fixation, that the dissolving above ground cannot master it. A water, which congeals, has at first a little crust, going on in it till it be quite congealed, but here it congeals from within to the outside, hence you may guess at that glorious foundation of projection, on Mercury of the body, making a natural, *stratum super stratum*, thus are the metals joined according to which the artificial work is ordered: we have a hint given how mercury of metals is clipped and allayed, and its *lubricum* is caught. Conceive not of this fixation, to be as when iron, is hardened to steel , and then reduced to a softness as Tin is of; this is called only a close hardness; which keeps the body in a malleableness, and keeps it so close together, that the fire above ground cannot hurt it, all hardness above ground may be mollified in fire, but not the other, because it holds all fiery trials: therefore as the hardness made above ground hardens bodies in the water, so on the other side, the water, which is in metalline bodies must be taken out, then it congeals. The subterranean air hardens the earth, earth remains earth, and turns not to stone, and the same keeps the water from running together, or congealing: keeping it from turning to pearls and precious stones, and such may be made of that water. To get the internal fire out of metals, though it be most high skill, however it is feasible, and found in its place, where I write of the like, in a more ample manner. I give a hint of it in this place, as Miners ought to do, of whose expressions I borrow now: The rest which wholly extracts this fire, which lies between the *project*, leaving nothing behind, that is, where the *Lubricum* and *Volatile* are together, leaves it, produces it, and excerns it. The *Mansfieldian-slate*, makes it appear, that its *Volatile* is gone, and its *Lubricium* also, where its impurity is yet between the *project*, and is not a fair pure work, but a compound one.

Chapter 7: Of the Ascension and Descension of Metals

This new kind or manner of speaking and writing of metals is caused by experience, for the first perpetual ingression of the *Ferch* increases and strengthens at first in the *officina* and *Matrix*, the *Mercury of bodies*, bringing it on to its perfect and full strength, being made wholly effectual and potent, then it begins by degrees to cloth itself with a body, at first he attracts and receives the meanest, which he puts off again in the first place, which is done the easier; for no body among them is sooner put off. For the body of *Saturn* is so thin, that it appears to the eye like as a fair body does through Lawn or Tiffeny; its spirituality appears through its body, its spiritual body is the metal of Mercury, or as I should rather call it, its proper, near, and special body, which work gives a manuduction unto many other fair works; for it makes a garment for *Saturn* out of the subtlest earth, after he rises higher, puts a harder and better garment on him, which is not easily put off as that of *Saturn*, or at least not with so small a work, which is caused by the work of the *Mercury* of bodies. For the *Mercury* of bodies by reason of its fluidness is the hottest, as he makes it appear in *Saturn's* ascension, putting cinereal body on him out of earth, hence is it why *Saturn* is so full of cinders inclining to a brittleness of ashes, and begins to sound by reason of the metal, though it be not very firm, however yet it is at the next place for incorporation: its sound is more deaf, is further off from Iron, and nearer unto mercury by reason of heat. Observe now at the ascending of this metal, it lays near the ashes, cleansed by the Saturnal water, but above ground it turns not to be glass out of the ashes out of salt or earth-water, or Saturnal-water, or out of sand or stone. But what is that pure subterranean Earth-glass, which if it sounds breaks not: it is a matter which Nature thrusts upon a heap together, which if you touch, it sounds, and is very clear, of a great compactness, and very firm, in this work it does mingle with ashes and salt water, and turns to a glass of earth, or to a dark glassy firm iron. Nay, tell me, if a metal or Earth-color, yea, a good sound metal be dissolved to a color, and is brought into a glass, does it not look of a copper-color? Yes truly. Therefore iron may soon be turned into another thing, which is done naturally, where such metalline iron color is reduced in *Hungaria* into a *Lixivium*, and is turned into a very good copper, however it retains the glassiness, though the color has exicated it somewhat through the mercury of bodies: for the liquid-ness it has still, and is nobilitated further to a malleableness and fixation, therefore take notice of this tincturing matter, which you find prepared by this body in this *Officina*, it reduces the iron to copper with abundance of profit .

Put these colors away and behold how the mercury of bodies is passed through many white bodies, and has still a fair white *fuligo*, and that very fixed, how finely is it clad in it, and makes a fair and pure body of *Luna*, into which he clads himself so strongly that it cannot be taken from thence by burning, because if passed seven times through the greatest subterranean heat, which destroys corruptible bodies, unless they be closely and compactedly incorporated to the mercury of bodies: nothing goes beyond that fire, neither of the upper, nether, or middle fires. Therefore behold how neatly Nature works and rises, calcining the whole body of *Luna*, which *calx* is nothing else but the body of *Sol*, its tincture and tinging quality it takes from the perfection and depth, which is in the fire, and can afford it, that color must keep so long till it descends again: there is nothing which can master this fire: the descending may soon be perceived by this ascending, and the difference of it is this, at the ascending it gets the tincture first, before it gets the body, but here it loses it sooner: and this is the reason why descending ores are more perfect, then the ascending ones.

Chapter 8: Of Respiring Metal, or of Quick Ore

Because with and by Mine-works an obstruction is made upon Nature, which is the cause that several metals are gotten, and distinctions put on them, that damages and losses might be the better avoided in the working of them. For as each received a particular name and property in or at the work, so in the digging of them, several manuals are invented for the finding of them and hereunto use was made of the Rod. To distinguish metals by colors is a curious skill, as Red gold glass, Myne green, black ore; however, their working is not so exactly known that way. That I may lose nothing in or at their melting, I use this means; first, I must certainly know the property of the ore, before it be beaten out, while it is yet underground in its breathing: for ores and metals breathe only underground, though they breathe in some sort above, yet the same breathing is very weak, not going far from the body; and the rod also sticks only upon the upper metals, which is the greatest advantage we have: for fire causes metals to breathe into a flame, and the fire-crates and *pit-diggers* cause only a volatility and closing compactness, where a threefold damage ensues. First, there flies away, not so much of the metal, but that also which flies turns to be volatile, and in the several meltings of it, always something goes amiss. Secondly, the remainder of it grows unmalleable, which hardly can be helped. Thirdly, elevated minerals are burnt to a compactness, which if not done, would prove very advantageous in the after-work, and chiefly they would be very useful in *Medicina*, being naturally prepared thereunto, which is the reason why many in their After-workings labor in vain, taking other improper minerals thereunto. For that *soffile Vitriolum* at *Goslar*, where neither silver nor lead grows in that Mineral, where it is prepared highly, copper may be made of it without any other addition: that Vitriol affords an oil also, which perfectly cures the Gout: if all these good qualities should be burnt away with the silver, were it not great damage? Therefore I take such a metal which attracts breath, and when the unbreathing or adhalation is stronger, then is it a living metal, because a lively quality is in it: for breathing things are alive, and breathing is compared unto life, such metal like a breath, proves as a child from ten years to ten years, even so this metal grows, till it come to its perfect state and body, afterward it gets another name, and consequently there must needs be observed a great difference in their working, and are asunder as much as a live thing differs from a dead one, which ought to be taken into consideration, because this distinctness being observed, affords a neat and pure work. Hither belong all mixed ores, which at separating are parted asunder and not before (as the usual custom is.) As in *Hungary* there is had everywhere gold-silver (that is, in it there is gold) which in its color and ponderosity is pure, has lost nothing, and is still in its working quality, and if it had not been interrupted, and digged up unseasonably, then that silver would have been turned to pure gold; that silver may easily be brought to a *Solar* perfection, and in itself is it better to be used for *pagament*, and is of better use for cementation. In like manner, the copper at *Mansfield* is good for it, and proves much better in the work, then other copper, for it wants but a little, it had been burnt quite into silver. The best quality of such copper is that they are of a deep color, they have not lost that, as usually *Electrum's* do. But this is to be noted, such ores are of that quality, that the bigger part of the body hides the lesser part in melting, it is not seen, nor felt, except at the washing and parting, there it is seen: while it is yet among the earth, it is a breathing ore, and is of such a compound, as you heard: you may confide in it without a proof, though the cake of it be of silver color or of a copper color. This is it apparent, how Nature augments a metalline body, protruding it upward from below, and that which is nethermost, is thrusted toward that which is uppermost, in a marvelous subtle way: for dead bodies bring still more to it, making it heaver in the mercury, then joins the nourishment also, which are the prepared minerals, affording their tincture, like good food, which breeds good blood: hence *flores mineralium* have their existence: if you cannot learn their off-spring in that way, there is no other nor better way for it. For this is the true tincture, and not that moldy or gross *album* or *rubem* so called, where such highly mingled ores or transparent Veins break, they may be cried up for an *Electrum*, but improperly called so; for the colors glitter so purely therein one among another, like in a *Chrysolithe* or transparent *Amber*.

Chapter 9: Of Expiring Metal, or of Dead Metal

Metals have their set time as well as all other creatures, they decay and die when their appointed time comes. For when Nature has brought the metalline body into *Sol* then by reason it wants nourishment, and is starving, then it comes down, gets a stronger exalting, and the attractive breathing turns to an expiration, an aerial breathing brings it to the fires-breath. If the *expiration* grows stronger in a metal than its *aspiration* is, then it descends by degrees, and decays, and then is called a dead ore or metal, for still one external body or other departs from it, at last in one place or other it makes a total *egression* with its breath, life, and seed. This breathing is known by the *particular Rod* of each; this also asks a particular place and work, because great gain is afforded both by this and by the living one. Consider it well, a metal, which descends from its perfection into another body, it is like when a man loses his lively color, and at last his body, that is, its ponderosity, and then the gold turns, not to a goldish silver, but to an *Electrum*, that is, to *Sol*, which has lost its tincture. This is a great piece of proof, to discern such silver-gold from true silver; in its gravity, it is found heavier than other silver, retaining the body, and losing only its color or tincture.

It is a greater skill to restore a tincture to a dead and decayed color, and to make it fix. At separating it keeps the quality of *Sol*. The like condition is red silver, in which has left its color, making an incorporation and union with copper, so that it quite dies in its body. To get this silver out of the copper, and restore its peculiar color, is a great skill, which Melters are quite ignorant of, belonging merely to the Chymick Art and its Laboratory. How many such *Electrums* are bought for silver and copper? The Buyers thereof have great gain in it. Of the same and the like condition are other metals of. Is it not so, all iron in *Hungary* is brittle, what is the reason? Because copper is in it: if that be gotten out in that artificial manner, as it ought, that iron proves so hard, that no steel is comparable to it? Out of that iron are made Turkish swords (sabels) mails, which no weapon or bullet can enter or break; these mails also are not very heavy.

Note, the *lubricum* at the descending of Mercury, must have room, from slippery things easily somewhat may be gotten sooner than from hard things: slippery things leave always somewhat behind; the same they do with their tinging bodies, putting them off still so, that in the ascending in their *Volatile* they assume the body, and elevate it.

Note, if you have any material in hand, and in your After-work you would know whether you must keep to the *Lubricum* or *Volatile* (these two must be your help.) Then your stuff must be prepared, either the slippery or volatile way; if you will have a body either ascend or descend, take notice of the flux in the metal: slippery ones are more open than hard ones. When tinctures begin to depart, which are a strengthener too, then the stuff grows more fluid, comes closer together than in the quick metal.

Chapter 10: Of Pure or Fine Metal

When a Metal, be it in the ascension or descension, is in its seven *Systems* or constitution, then it rests or endures, till it comes into another body. If you meet with such ore, it yields the purest metal, that may be had in the world: our Melters call this Super-fine. But our Super-fining, which hitherto was in use, is an impure work in regard of this: for in the clarifying, if it misses but the least grain, then is it not yet right. Such metal, as may be easily conceived, is pure good and malleable, loses nothing in any work whatsoever: though all metals may be made super-fine, yet none can be made finer than gold is, which no element is able to touch, to take anything from it, or to turn it to a *Glimmer (Spolium)* or cats-silver, of a glittering quality.

Silver at *Marychurch* at *Lorrayne* is more fine than others. Super-fine is called that, when a metal is pure, and rid of its excrements or dross, which may easily be taken off, and hinders it not in its fining. In silver Mine-works there are often found such natural proofs of pure and fine ore, that it might speedily be digged and broken, though it must be melted again by reason of its *Spolium*, or reason of strange colors and flowers it has robbed, yet it easily may be performed, which serves afterward for an instruction, how Mineral-colors must be obtained, as *Azur*, *Chrysocolle*, though they stand in the Mineral-glass: such colors love to be in such pure ore but are not so soon inoculated, unless it be in the *Sude* or coction, in which the metal is very pure, and yields more naturally the mercury of the body, be it in the ascension, or descension, assumes then another body. Hence it is apparent, how the same ought to be proceeded within the artificial After-work, out of one body into another, how the body, in which it is, and from which it must be had, ought to be prepared, namely, it must be made pure and Super-fine. It appears in the *Italian* gold, especially in that of *Wallachia*, in which it is most pure: how that mercury of metals puts off his body, and the mercury of the body comes from the mercury of the metal, puts the gold together into a close body and Regulus, and it is seen in the gilding, how firmly and closely it sticks, wanting but a small matter of an augmenting quality, its *Spolium* is only obstructive thereunto: it is of a transmuting and elevating quality, if the other body be awakened also: for a body which is between awaking and sleeping effects nothing, it must be awakened wholly: if at the on boiling of a metal, as of that *Italian* gold, be but the least impurity, that is, a heterogeneal part, it could not be brought to a compactness, which is seen at gilding. Therefore, you must give an exact attention to learn to understand what the *prima materia* of metalline bodies is, and how their *Elevation* is either obstructed or augmented; how homogeneal things are brought to a body. It is apparent in the mercury of metals, how close and compact it stands together in the flux, which flux cannot be taken from it: purity is the reason or cause of that compactness, being there is no other metal mixed with it: as soon as any metalline body joins with it, then is it disjoined, be it what metal or body it will. Hence it does appear, how metals are brought to rest from their labor, namely, if they be first pure; for into pure matter may be brought what is intended for it; which appears in the mercury of metals, its purity is the cause why it does not appear to the eye, but only in its flux or hardness. The mercury of metals is the flux of the mercury of bodies, that is, when water comes to it, or the mercury of metallic bodies is come into the water instead of the air, which otherwise is in the water: take it into consideration, what manner of skill is required to get wind or air out of the water, and bring another mercury into that place, if you get the air, which is in the earth, out of that earth, and in its lieu you get the mercury of metalline bodies, then you have a Mercury *in Coagulato*; endeavor how now you may coagulate it, but not in the ordinary, common, and vulgarly known way. Bring still another mercury of bodies instead of the Marin-water into that water, then you have a fair pearl, take that same mercury of bodies, reduce it to an earth, which must be pure, instead of the air then you have a pure jewel as pure as may be had from that, earth is in its color, or you may put one to it, which you please, it is a thing feasible. These and the like pieces are afforded by purity, all which the work of Nature is a leader unto. (Men that cast so many foul aspersions upon Philosophers are unworthy and not to be regarded) nor credited, what they can foam against their rare and glorious inventions about the three principles, from whence all these things have their Original. Make trials of it, you will affirm to be true, what I have said.

Chapter 11: Of Impure Metals

There is a found store of metalline ores, but few of them are pure, and few there are that break or grow one by the other; therefore these must be separated and spoken of apart: The great work and expenses which their cleansing requires from their grossness let Melters speak of: separating has been invented, at which some good things of the ores do stay, the rest flies away, and their fining is useful, especially when ores or metals are in their ascension, though it be changeable. But to find *Electrums*, and bring them to good by separating, differs from the former fining, and requires a singular way of melting. Cunning and subtle Artists may pretend to get silver out of iron, (I believe they may, if there be any in it) as they do in *Sweedland*. *Osemund* always contains silver, which is only off driven, and calcine away the iron, and thus they cheat people: Can they do the like with the iron that breaks in *Styria?* No such matter. Therefore, take heed of such cheaters, and take notice that nature loves to keep her own ways orderly, and keeps together two and sometimes three sorts of ores in their ascension and descension, whereby she intimates a way unto the After-work, but men in their fancy think upon other means, though to no purpose. View all the Mines which are in *Europe*, you will find no other ores but impure ones, that is, a mixture of them, for their nature makes them, as much as I ever could learn: if you can show me the contrary, I will assent unto you, and this is the second Argument, that metals are in their ascension and descension into perfectness: if each had its peculiar work and instrument, then men needed not to take so much pains in melting to separate them. For it is a difficult work to join weeds and stones, because these are heterogeneal, and are of differing matters: but the other joining soon together, requires special working to be separated; therefore, view exactly the bodies two manner of ways, which is no small instruction. First, in what manner you separate the ashes from the fumes or food, this ministers already a twofold separating of metal, the earth from the tincture, there you have a twofold separating, and so forth. Secondly, take notice of the Flux, to drive the cold fire with the warm, and the warm with the cold, then you will be able to separate the bodies from

Mercuries, then you have already separated the metals without loss and damage, use thy self exactly to it, and be careful in observing their names, not regarding the miners expressions and terms, for the names they give to ores are false: for those, which build and dig after clefts and passages, have their names of their bodily matters after the sorts of the minerals, and are distinct therein. But you must call them after the sorts of the seven bodies, and learn to prepare them; this work is of greater utility. Men are at great expenses to get corrosive waters, to get asunder these metals, they do it also by way of melting and casting, but such waters add great poison to the work: it is a better way to do it with *Lixiviums* or sharp waters, which are not so poisonous, learn their preparation. There is another kind of impure ore, of which I made mention here and there, which contain Mine-slacks, you may read of in the *Chapter* of the Cinders, but there is a difference between the cinder and the slack, for slack are more corny, yet that also turns at last to cinders. These slacks are the cause of the cold fire, and cinders cause the warm fire, or the *Vredines metallorum*; those cold slacks are hardly gotten off from the metal , because they come from the cold flux fire of mercuries altitude: for as the cinder comes from bodies; so are these slacks of mercury; it is seen, when you will have slacks of other matters then usually they belong thereunto; then the mercury of bodies is roused, which by the work is nothing else but a closure and stream; for if you can conveniently get away the slacks, then you may perform and accomplish something else with the fire of mercury, and it is neither usual nor artificial to deal much in cold fire: some miners call it *Mispuckel*, *Nodus aeris*, that Latin name they put upon: it is true, it is very hard knit together, it is difficult to dig it , and to make its ore to good, and Antimony also is hardly gotten from other metals (except from *Sol*) without damage; however, with advantage it may be done curiously, only you must be expert in *Antimonies* qualities. For they belong together and are joined, as Tin and Lead, *Wismuth* or *Magnesia* among or between Iron and Copper. This is a good direction, and is sufficient for such, who know what belongs to melting.

Chapter 12: Of Perfect Metal

Who could tell what gold and silver were, if they were not known in their perfection, for when they are perfect, it appears, when they have their color, their weight, their malleableness, their flux and hardness: and this perfect metal Nature has produced compactly and purely; for such perfect, pure, and compact gold is found in *Hungary*, in the white marble, which presently may be broken, as also silver, copper: the difference between the perfect and the pure is because metals are not pure before they are perfect; and so there may be a perfect ore, which is not pure; which defect is found in many other metals, which come to their perfection as soon as in any other Foreign parts, but in their perfect purity they are defective sometimes. This is to be noted by this metal, a body must first be perfect, or brought to its perfection, before it can be fixed, and it is of great concernment to know rightly what fixedness means.

A body which has its due tincture, weight, and graduation, yet it has mixed other obstructive impurities, here comes the work, and nature begins to copulate these two, tincture and gradation brings the metal into a purifying; this purifying is fixation: for pure is as much fix. And note, that the ground of the first is the body, which is a secret, into that I must bring the tincture and gradation as well as I can, and take the tincture from *Sol*, which is a thing feasible: then is it an *Electrum*, which is a water; for in water it abides, then I take its ponderosity for it, and bring it into an *Oleum*, or into a *Sulphur*, the body remains still, for in the ascension mercury lays the foundation of the body, as an *Embrion*, to it comes ponderosity, which makes it formal, then comes *Lubricum*, after that comes the *Volatile* with the tincture, and perfects all the rest it has need of to its perfection. Why does reason play the fool in despising the ways of nature, not observing her course? For behold how wonderfully she brings redness into copper, turning it into brass, but is not fixed, because it was not her intent; it is a mere color which all other ores easily embrace, but is not fixed, which color is easily driven away with wood and coal fire. Therefore it is a thing of great concernment to learn rightly to know the bodies; for at dissolutions the property of a pure metal is known, what its tincture, body, salt, and ponderosity is, especially if exactly be considered the anatomy of all bodies after the Chymick way, how curiously and properly are they anatomized: we call the immature spirit, a spirit of mercury; the perfect tincture we call an *Anima*, or Sulphur, the ponderosity is called the salt or body, for the After-work confirms the same, that that fixation does not only hold in the fire, and all corrosive waters, especially that of *Saturn*, which is a precious one, more than other Aqua-fortis, but better in the malleableness without the *Quart*; it holds also in the cement, because it comes out of it. Therefore, it is to be admired, why men talk so strangely of it, when they know nothing of it, from whence it is, or what the cause of it is! But it is so, the one harkens to the tale of the other, and know no more than they have from hear-saying; for they know not what to say, nor how to help what the work ails, if out of tune and disordered; and if any fix and lasting medicine is to be made, then that metal is best, even as a vegetable which is come to its maturity. This process must be observed, else all labor is in vain. For how can you destroy the body of a thing that has no body, much less can you have the tincture of it, before it comes into the body: a tincture may be gotten from it, but not all the colors of that tincture: here exactness must be used, because it is the greatest skill to do so. One thing more I must needs speak of, those that aim merely with a great and deep longing at riches should view God's mysteries everywhere, if they compare the Scriptures rightly, will find an instruction, that a spirituality is held forth in an earthly thing: if a metal be brought out of the earth, and out of its *Officina*, like a man that is set into another life, it abides and lives without food, is not dead, but is alive; though it does not act, yet it rests, and may soon be awakened: thus we hope also, that in the other eternal life, where there is *Sabbathum & Sabbotho*, things will be done in a more glorious way. God makes use of us here for his praise, to make use of metals in that kind: it will not be for our misuse and disgrace, but rather put greater honor upon us, more than ever we put upon metals.

Chapter 13: Of Imperfect Metal

The imperfect metal is the wildest among all the rest, because it contains all impurities with a confused mixture. Usually there is set in a work of many pieces, which no man yet knows what they are, in which you find matters joined, which are prepared and unprepared of many bodies. Now if you will reduce that matter and cleanse its metal, then the unprepared matter must first be washed away, and regard not the matter so much as the metal, which is yet tender and young in it, you must not calcine it, but another fire belongs to it. This perfection consists in nine several pieces, which must be well considered; each must be handled as it ought; if so be, the metal shall not suffer any wrong and damage.

First, consider well, whether the ore be in ascension or descension, then you may help its color and whole body: for that metal, which is in its ascension must be aided in its *folium*, but the metal, which is in its descension must be helped with *Spolium*, else it dies away, because it is not perfected in its due place and *Officina*, and that metal is in a form accordingly. Thus, make your proof, and be in your After-work the more encouraged.

Secondly, you must take notice of the nourishment or food, which as I may say, is not fully digested, further the same in its concoction, else the excrements cannot well be gotten off, which would be the cause of unmalleableness

Thirdly, you must take notice of the flux, that which is fluid must be fried: for if both fluxes should be opened, they would hardly be congealed again. For to dissolve a new running mercury of bodies, which *per se* are running, would prove a mere dust and atom.

Fourthly, remember the hardness or fixation of it, in what degree of ascension or descension it is, that you nay add thereunto an equal fire of its body: for cinders require a sort of fire, and *fuligines* requires another kind of fire, and calcination must have a due fire: then your proceedings will be right, if you know the proper condition of such bodies, else you wrong them, by making it brittle, and flies off at a second casting.

Fifthly, make a true distinction between the two imperfections: from thence this metal has its name; the one is of the body, the other is of the instrument: first, take in hand the instrument, and endeavor to help the body in the preparation, and stay its flux or running into another, and you drive the *Schwaden*, or the cuticle of the seed; then goes the cuticle of the seed, and the *Ferch* or life.

Sixthly, consider that fire, which nature has, that you do not encroach or entrench too far upon the bodies: direct and order all your work upon the *Uredines*, or soap, join it with your fire to the instrument of the body.

Seventhly, you ought to be instructed well about the tincture of the *Electrums*, that you put not of *Sol* instead of *Luna*, because you are not experienced enough in the sleeping tincture or color, which you are to awaken in the descension, or when in the ascension, you are to strengthen and to keep them. Painters have a term in their profession which they call elevating and shadowing, the same you must apply here to bodies, when they are in their perfection: therefore, anatomize the bodies.

Eighthly, we see Limmers to mingle their colors with water and oil, and so carry them on, you must learn a due preparation of the water, to imbibe the color, which is a metalline water, whereby you corrode with not destroying the colors, which you intend to have, if you make use of any other water besides this, then all your labor is in vain; cool with oil , then it remains pure, and thus you have much strengthened the *Folium*.

Lastly, take notice that the metal be strengthened so, that it looks for no more food. All such things belonging thereunto you find enough by this instrument, or not far off from it; or else you may bring it easily thereunto: all will be but imperfect, therefore put your help to it, you have natures half work for advantage, with great pity it has often been beheld, how such noble precious instruments have been consumed in fire, which could have been for several good uses, the workmen in their carelessness not regarding them. All other metals belong unto this: for very seldom is found a whole perfect metal, which should not want one help or other. He that is acquainted with these will perform these works with greater advantage and more utility. For there must be extant a perfectness, if anything shall be purged, so that another pure or fix thing be brought in, be it the color or ponderosity.

Chapter 14: Of the Soap-metal
Metallum Uredinum

If our upper Elements could not make a metal to be manifest and apparent to us, who would have undertaken to work any metal. The great heat and cold which is under the ground is the cause of the body of metal, according to the predominance of that heat and cold which is in the body. The deeper the heat or cold is in that body, the fairer is its tincture; this is a true saying. But what this heat or cold is above, and what is attributed to the Planet above, understanding men must know it.

At first when the *Ferch* goes forth, and goes along through the earth with the seed in its firmamental *officina*, then it comes forth sometimes, as far as its breathing may reach before it be infringed or allayed; it still carries some of the purest metal, and the superior elements afford their virtue for their joining: for where the elements are in conjunction, there they make a body, and force something from the *Ferch*, then it rises and falls, as it falls, so it lies still, this is the reason why the corns are so strangely fashioned, they are either round or oval, and so come to the metal accordingly, and is found: sometimes it falls into the water, where it was caught and overtaken, but most of all it loves to be in the *Uredines*, or soap, hence got that name. *Soaps* are mineral springs, where the metal loves to lay, these spring from below, and these are found sometimes wholly gilded over, and they cast upward taps of such color. In this Country, such springs are near Rivers, at the foot of Mountains, but in the *Almen* they are on the top of the soap as it were upside down, on these stick the *Ferch* which passes by it, or the air which forces it from it. The whole business lies in the metalline waters as they are dissolved in their salts, thus they easily embrace the *Ferch*, or the metal of that *Ferch* and seed which it carries forth, and that goes for the best Tin which is on the soap, being of an iron shot, which iron is of several different degrees, but here in this place it is not energetical, as being obstructed by two parties.

First, the water is an obstruction unto it, for ore must give way to water, it is carried away: such soap ore is seldom found by the *Uredines*, or in the soap.

Secondly, this *Scobes* powder or dust is not in its right *officina*, because it is not in this subterranean air and element, where the singular *officina* of metal is, and in this manner of condition are almost all the soap-metals in *Europe*, but in other Countries where there are none, or very few of these nether metals, if there be found any of them in the Dam earth of the highest or uppermost metal, then is it another matter, needs not to be related here, because it serves not for our work. Such metals are the best, because they lay deepest at the metal; and if you entrench too nigh unto them with the fire, then they fly away with the *Ferch*, by reason of their transcendent superfineness: therefore when this ponderous *Ferch* and seed is grown very hot in both heats of his work, especially in the descension, then is it mightily strengthened, in the consuming it grows more homogeneal to the upper Mercuries of the earth and its instrument, therefore it does associate thereunto, and obtains a going again, and this is the ground of ascending and going upward, besides or without of the *officina*, as much as is known.

Chapter 15: Of Inhalation or Inbreathing

It is a curious work to build after-clefts and passages, for it seems a thing to be credited, that in them the breathings have more their going in and out than through the whole earth besides, but it is not advisable to take such fetches about, because there is a nearer way unto it, namely, unto the metals, as if one would go to the well, and goes all the streams down, through and by all the springs, whereas there is a nearer way to go the foot-path, the same condition it has with the halations and breathings, because the fumes and vapors lie therein, and are obstructive rather to the breathing, than helpful. Therefore, look not so much upon clefts and passages, but rather upon the halations or breathings, which keep their stroak throughout the earth, because the same does not hinder its course, as men may deem. For we must know that the ores do breathe only underground, and these breathings are orderly distinguished. This is a remarkable thing, that ores breathe from below where they are, and this is the difference between the halation, vapor, fume, or breath, which goes oblique, the cross way, lateral way, or afar off. Therefore here is to be spoken of that inhalation, when it goes in its own strength, what it shows and signifies, the fume is lent unto the metal from the superior *Solar* rays, from thence she may have what she will, and what she has need of for her sustentation, that brings the nourishment, and drives all downward to the metal and the lower planets, but no farther than upon the earth, but through the Dam earth *per cutera terrae*, there the ore or metal receives the inhalation, which arises from the work, makes use of it: for it is manifest that elements cannot reach deep into the ground. Seeing the halations are invisible, whose being cannot be seen in them, a man's breath can sooner be discerned and seen, than the halation may; which may be known by a *Rotten,* when a metal works, then it breathes, which breath it draws again, and so forth: then the Sun must come in to help, for if the *Systeme* is at a inhalation, which must be known artificially, there is the purest earth, from whence man is taken, as some ancient writers do conceive, which earth contains many hidden virtues in the After-work of metals. This inhalation serves chiefly for that purpose as you heard, to draw, swallow, and take down the food into the body. For you must not think, that she lets it lie in the body, after she has brought it in, but it is a *Visretentiva* to keep that fast, which it drunk in, and to that end she prepares in that whole journey from that place where she received or borrowed of the Sun, which drives down all ponderous things, as we see by all such juices and saps which are driven downward; in this drawing down it has this quality, what saps or juices it takes, the same sink down deeper more and more, and sublimes them the more highly into its *flores*. But this I call *flores*, when it cannot sublime any further, and brings it at last to a garment or body of a metal.

He that knows these flowers rightly, he has learned something of nature, and he that knows of what they are made, he is come yet further; but such flowers are made naturally of mean stuff, which works by a threefold fire, among which the inhalations fire is not the meanest, but the chiefest, because it is the next unto perfection. But this breathing is more a fiery aerial heat, then it has of any fire, because it flames without any kindling, otherwise it would combine the nutriment in the *officina*, she conceals rather, keeping things together, and strengthens them, it expels humidities and vapors; and consequently, it causes those evil vapors, fumes, and breaths, which poison the whole earth below, and are the cause of many dangerous diseases, as it has been known.

Chapter 16: Of Exhalations or Out-breathings

Seeing, stirring, and moving, must be continually at the metalline work as long as the metal attracts, and especially the in moving or taking into, hence the next inhalation of the living ore must be stronger, therefore Nature affords so, that because the halations can come no further in the metal, because metal is the end of the work underground, that then the exhalations must draw on, and must turn into the metal, the return of that metals breathing from below into a true natural breathing of metals grows so strong, that it kindles, yet without a light or flame, but glows without a flame or light, and purges that which is to be purged in the work, not only in the work in the *officina*, but in the whole matter, till she heaves and protrudes the pure atoms of the earth, which the Sun gloriously dissolves again into juices; for the atoms are the flowers of the terrene-salt, she cannot endure them, neither are they useful in the earth, but must be dissolved again in the upper air; but this exhalation, though it be always about the metal, yet is it apparent in its *Rotten* or *Rubedo*, what the ore does, then the breathing seizes on the *Rotten* or *Rubedo*, and kindles them, then is the metal in its decay or decrease, and is a dead metal, and most of the fire goes forth, which kindles this *Red*, and the cold stays by the nether metal, which grows predominant, hence we see what *Uredines* are able to do, when they are predominant; in metals, then the cold gets the upper hand, and disgraces the ore, bringing it as low as lead; but if heat get the upper hand, then the metal grows higher and higher, and more lively. This is the difference between the heat of fires, which is called *Uredines*, the vehicle of heat is the instrument or matter which affords the place of it. For *Uredines* are the cause of something, but fire destroys it, and the *Uredines* may be without the fire, and above the fire. The most ancient philosophers, call it *coelstes & infernales vires*; for it is apparent in gold, that fire cannot prevail against it, but only against the flux of it, and cannot consume it, and the Mercury cannot hurt either, only he brings hardness into it, which is good, but to what purpose? For he is not turned to any silver; both these breathings, upper and nether, in and out-breathing comes all to one, and is not only its quality to show and make known the metal and ore, but passes through the earth also, to keep them from turning into stones, to stay and keep them porous one upon another, and to make pure: for it purges the earth, as the *Solar* breathing above does the air, and as the same breathing above brings and causes wind, mist, rain, hoar frost, snow, even so the subterranean breathings causes vapors, fumes, breaths, saps, minerals, soaps, etc. and brings forth gold out of their seed. He that understands these nether, upper, and other metors[7], he may make all manner of quick After-works. For Art treads in the foot-steps of Nature.

---

[7] This is how the original text was printed. -pnw

Chapter 17: Of Corruscation
Or of Adhalation, or to Breathing or Glittering

Miners seeing this breathing, ascend by night like a flame, they count that only a breathing and none else, which *per se* is not a true breathing but only a *corruscation*, for it burns away the excrements, not wholly, only such things that are naught and poisonous, purges the air from the same stink; for where should the cuticle of the seed get out, if it should not be carried forth, partly by this flame of fire, and partly by the water? For this stink weakens the ore mightily, especially it spoils its color, whereas it may easily be burned or washed away, before it grows to such a stink, be it in *fossils* or liquors. This breathing or glistering strengthens the cinders, it touches them not in a way of heat, but in a way of fire, and is not an up-boiling, but a burning unto.

Seeing this stuff which this corruscation seizes on, is found often in a place where no metals are, therefore is it very deceitful, however commonly and for the most part of a metalline breath. For note the metal never affords or casts any flame, neither does it consume in burning, but flies away and vanishes, you cannot know which way it is gone; therefore these *exuviae spermatis*, or husks of the seed, is a greater infection, and plainer evidence, that there is an ore at hand; because it makes many glorious things apparent, which the fire seizes on, it must not quite be melted from, its outgoing sign that it does no good by the metals, neither can it stay with it, unless it stick close. Here you may observe again, that she does not good to metals, but a warmth that does not come from fire, as a natural heat is in man without fire; for where are the coals which Nature uses for ores, yet is it hot? Where is there any better blast to make things congeal? A cold without frost is another matter, as it appears in Pearls. I call it *Uredines*, as a warmth without a heat, yea, this heat and frost causes a food unto fire, it cannot endure it anywhere about it, therefore if you bring a metal into its heat and cold, then you have already a ground for the After-work; you have little learning or skill, if you do not know so much.

Chapter 18: Of Folium and Spolium
Or of the Schimmer and Glimmer

The second or other thing which apparently shows and distinguishes metals, is the color. The condition of color is this, that they are not so discernable, by reason of their darkness, in which they naturally are, they are elevated and brought to a *Folium*, or are exalted through a *Folium*, as all obscure and non-transparent *folium* does the same in a *lucidum & translucidum corpus*, the same must a transparent *folium* do in a thick body, as metals are an *opacum corpus*, shining like the moon with a borrowed light into the body, and the *folium* gives it forth out of that body.

Such a *folium* or glitter may be made artificially, but Nature shows that it may be had from a *Volatile*; though it be true, that that *Volatile* may not be bad, unless in its seed and body; yet it is soon extant with the body, for he that knows not the condition of the *folium*, he can neither in the flux nor congelation bring any ore or metal into its true tincture: the condition of this *Folium* is, that it is as thin as any leaf in the world can be, of these leaves is composed a metalline body.

Counterfeit Chymists busy themselves very much wi.th their *Stratums*, *Superstratums*, but here it is naturally thinner than beaten gold, and this is called *opaca perspicuita*, as in brought leaf, not gilded in, or silvered. Here lies a curious skill for tinging, with this leaf, if brought into the *Glimmer*; one knowledge is the cause of another, and as it is seen in a glowing and glittering splendor, that there is no *exuviae spermatis*, or hull of the seed at hand, neither may it be known in any *folium* or *spolium*, by its *speculum's* or night lights, to prepare that leaf and *spolium*, which soon does loosen and free itself, and this is the chiefest thing, for it goes off with the worst poison, and comes on also, and must settle and rest.

Chapter 19: Of Fuliginous Vapors and Ashes

Touching the white soot of metals, which is a precious garment to silver, if only brought into Mercuries, wonderful to behold, that the corruption of metals and of the first bodies should afford the highest and the uppermost turn to be the lowermost, where these soots are found surely there is metal at hand also; but fair and more natural is hardly seen, then there is in *Styria* where they make good steel of it. In *Wallachia*, it breaks or is found near silver, and this is the surest thing in the metalline work, as well in the Natural, as in the After-work, which seems most incredible unto the people; and again, that which is most uncertain proves to be most certain; the reason of it is, because they know and understand it not better. Miners have in their clefts and passages things hanging and lying, in which great matters might be had and found, but require great toil, and is a dangerous work by reason of their poison. Though the Halation and breathing goes along, yet it carries many things with it, which hardly may be gotten from it, neither does it quit the pain and expenses. Miners call the matter after their way indiscreetly and inconsiderately, as the greedy Alchemists suppose, that when men speak of the Mercury of bodies, to be the Mercury of metals; or the salts of bodies, to be a salt like other common salt; Ashes of *Saturn* are seen here neatly, before they come or go any higher, and before they do turn into silver, for soon do they embrace *Antimony*; the same the congealed water or coagulated *Saturn* has likewise, and is a pure proof to all metals, and grows deep.

Then there is a pure Wismuth, which is gross, contains a congealed water of *Saturn*, which is found also with great gain about metals, when they are reduced to their first matter, then it ascends into a glass metalline work and the ash-work ceases. Chymists in old times, and now also made themselves very busy with their salt, to make all bodies potable thereby, having reduced them first into salts: but here is an *album* out of ashes, which ashes afford salts, which is but in vain; for ashes are garments, intimating and pointing at the thing which is clad, and the ore clothes itself with it, when it is coming near the day. The condition of *Potabilia* is otherwise, they must be brought into potableness, and is done in a clean contrary way then they go to work with. But these are served well enough, that go for nothing but for gold, Nature gives to such these garments; it gives the slip before men are aware of, I mean the body, therefore look well unto bodies.

Chapter 20: Of Scobes and Metalline Water
Of the Schlich and Lye (Lixivium)

Nature in her work must still have an increase and decrease, some things are coming, and others are going; and as above ground at the melting, and at the hammer metals do lose somewhat, so is it under ground. But how these subterranean decreases are discerned, which like crumbs of bread should be preserved. Nature being provident keeps them together also, bringing them into the day light, that they might be brought to good, and that metal is known also to be under that ground by the *Scobs* or *Alkali* which it excerns, namely, the rocks, stones, flints, which sucked nature dry in her work, these are the offal, if empty of metals, and if some good ore be in it, then that may well be called *Schlich* or sliding, because it slides away from the work, it steals away, where such sliding is, take notice of it, for it breaks off from the matter in the *officina*, wherein metals are in their work, and perform their office, and that *Schlich* or sliding is a sure sign that metals are in that place.

So, the *Lixivium* also, or metalline water, is a sure forerunner of metals, and it dissolves still in the work, and loosens somewhat from the metals, in which there is a *Schnede* and virtue: for as I have said, when I made mention of minerals, where there are *fossilia mineralia*, there is sure a Mine-work. And where such minerals are, they soon dissolve into a water of *Lixivium*; and this is the difference between the water and the *Lixivium*; water affords only the mineral, and the metals allow the flowers thereunto, out of these comes a *Lixivium*: the effect of this water and *Lixivium* is well known, for they carry in a hidden way flowers or tinctures of ores. The Cement or *Lixivium* at *Schwelnitz* in *Hungary* corrodes iron into a *Schlich* or sliding; and if that iron sliding is taken again out of the Dray, and is cast, then is it good Mercury. There are more of such *Lixiviums*, but men regard them not, that water also is a sign of a very pure mineral, for let us consider the water at *Goslar*, does it not yield pure white and red Vitriol? And in several other places there is found good copper, silver, and lead: out of these waters may be had again minerals, as you please or intend to have them, without any great pains-taking. For if a mineral is exsiccated, then it has no more the strength to run forth, but grows tough, and is dried up sinking into the ground. There are waters in *Hungary*, favoring of Sulphur and Allome, which affords store of gold and copper; alluminous water in *Misnia* afford silver and copper; the waters in *Bohemia* which taste of saliter or iron, afford several sorts of metal, except gold. The Mines in *Styria* have their special metalline waters and *lixiviums*, besides other excellent Mine-works, insomuch that the whole mines are of mere steel, copper, gold, silver, quicksilver, and other things men's hearts may wish for. The salt waters at *Frankenhouse*, do they not signify that there are curious Mine-works there? Which if the inhabitant took more notice of, and were more known, what gains could they not make of them? The more these waters are running, the better they serve for many uses; if they are standing, it is a sign that there are evil and bad minerals at hand? And that there are cide, matters, and minerals that were gone and left behind in abundance, of sediments and fumes. Take heed of these.

Chapter 21: Of the Scoria and Exuvium
Of the Seed and Hull of the Seed

The utmost, and last decrease, and diminution of metals is cinder, which is very good and harmless. I mean that cinder which metals put off by their *Uredines* or soaps. For the *exuvium* or husk which the coruscation or by-breathing leaves behind, and is like unto a cinder, is another sort of cinder, like unto that which comes from the forges of Smiths and Melters; for besides that they weigh their bodies, they cleanse them also, though the lye does purge the Minerals and colors, yet themselves also do purge, as is seen by the cinders which every metal leaves behind in the fire, which however are not like unto this. By this cinder, it is seen also, that there are metals at hand, for the cinders of Minerals, which the coruscation causes, are palpable, known, and visible: understanding Miners know the same; there is a metalline cinder among the slacks, but not known, which is the reason why slacks are so brittle, else they would be malleable, or else they could be cast compact; some have undertaken that work, but they could not get off these cinders.

Touching the *Schwaden* or *Husks*, these do fairly intimate the departing of the seed, and of the work of all bodies: for when the hull ceases to work, and has no more food, neither of the minerals, nor of the bodies, and now the *Folium* is gone into *Spolium*, then it is at separating, and breaks the band of the leaf and seed, which is called the *Schwaden* or *Husk*, it is an extreme poison, it destroys all that lives, especially breathing things; for it is its intent to return thither again; therefore to what place so ever it comes, finding things that move and stir, the same it destroys, and itself also; and at last, it returns to the *officinea* of the seed, helps to glue on and imbibe, and turns to be a band again. Where such poison is, be sure that there was silver and metal there, but never in that place, where it went forth, and is turned into another thing; for after the decay of each body, and of its leaf, and after the off-plucking a singular husk, the last it makes is the fiercest; for there it lies like a *Speculum* upon the water, and easily turns into nourishments, which is the reason why it turns so soon to the metals, and so the nourishments are infected contrary to their quality, that it wanders so up and down is the reason, because it contains some of the seed and of the life, and is the untowardness: for the seed must have something in which it may lie, which if it be not one of the bodies of the seven metals, then is it such a poison or husk, this is the skin wherein it contains itself, so long till the coruscation forces it away, which then is death unto it.

Chapter 22: Of the Shining Rod, or of the Fire Rod

He that intends to meddle with rods, must not follow after his own fancy, nor bring in novelties unto Mine-works out of his suppositions. For nature endures not to be curbed in her order, but men must be regulated according to Nature. Concerning the rod, good notice must be taken of the Breathing, and this *Lucens virgula*, or fire-rod is ordered upon the operative and attractive breathing; for if it were strong, though it does not kindle, yet it does its office, through and by a heat appropriated to its quality. A great heat in a furnace puts out a small heat, light, or fire, the same effect has this breathing upon this Rod, which being kindled and stuck in, is put out, no upper air can hurt it, for our upper fire cannot live or burn underground, for if a light or candle be hit against a stone or earth, it is put out, because it cannot fall in. It attracts the nutriment, which makes this Rod burn, and sucks it dry. This is remarkable in the matter of this Rod, that it has an unctuosity that does not burn, as the seeds husk is, whose poison put out lights which under does ground in an extraordinary manner. And the breathing above ground does the same in a peculiar manner. Few miners know this fire, and is the only means whereby the inhalation is known. This fire-work, thus extracted, is of special use in Mines, and then in the After-work is of several uses for metals; of the which more in its due place: to descry fire by fire is no mean skill, and the subterranean fire can in no other way be mastered. Ancient Philosophers have written much of it: how superior elements make their juncture with the nether, intimating also that by the means of the middle, Elements must be learned the emergy[8] of the superior and subterranean. For they are spirits which join the souls above the grosser bodies below, else nothing could have any success within the earth, and for their operation there must be a *medium*, a gluten: to tie and bind fire with fire, is a strange maxim, but is a true one: hence comes a reserve of fire, which never burns; all three must be together: for the upper is the light fire, all the metal is flame-fire, and the nether is

---

[8] Exact spelling from 1671 edition. -pnw

the burning fire. This we shall know on the great day in the other life, where God will separate the burning from the flame, when it flames; then will the burning consume Hell, and Hell will soon be burned away, neither will there be any clearness, but darkness, because God bestows the light upon his chosen ones, which is neither a flame nor a burning; here these must stand together. Be acquainted with its friendliness and friendship, make it your advantage, which is abstrusely hid, and goes invisibly.

Chapter 23: Of the Glowing Rod

Because the stirring of the Rod is fallen into abuse among many people; however, it is a fundamental way to know and to bring forth the metals, if duly and naturally used. It is an undeniable truth, as I made mention of it afore, that metals do breathe, and the same breathing is invisible; the best means to learn it is the Rod, this is the reason why I call this rod the glowing rod, because it reveals and shows the exhalations of metals, which exhalation is a fiery heat, though it casts neither flame nor sparks, yet it is of such an heat, that it glows in its manner; and brings the rod into a glowing, which is an infallible sign that there is a living metal at hand. This rod, if it should be more glowing without a flame, there must be used special matter which receives that fire, else it cannot be done with any utility. Concerning that rod; it is a stick or staff, of the length of half an ell, of hard wood, as of oak, unto it is fastened the matter, even as a torch or link is made of pitch or wax: this matter must be of an unctuous matter, either of an animal, vegetable or somewhat else, which is upon or above ground; it must not be too strong, otherwise it sets the breathing on fire; no talk, wax, pitch, or resin, is fit to be used, nor any mineral, otherwise that breath, mineral or metal would set its food on fire, and consume it; it must be a calx of earth, which catches that heat suddenly, and smothers in a heat as calx does above ground, being moistened, it falls thus hot off from the rod. This lime, burnt above by breathing is good for several uses. But this breathing does not seize on the unctuosity or Mine-wax; else how could the nourishment of minerals prosper? Miners call it a *Spath*, a true calx of the earth. It is not corroded and seized on underground, because it has too much of humidity. Some call it a Mine *Mermel* or subterranean *Mermel*, but is no such matter, because he does not endure the weather above ground, exsiccated, and by breathing is kindled at last.

The huge Mountains in *Norway*, the ores in *Swedeland*, seize on it and corrode it, they are full of *Spath* or calx, in their glowing they grow hollow; for if they were in a flame, that land had been consumed long ago. For as soon as that Earth-water comes forth from this *Spath*, and the exsiccating earth comes forth also, then the hidden fire falls into their places.

Chapter 24: Of the Leaping Rod

While all metal is in its purification, that it neither rises nor moves, then it has its breathing the same as it is of a singular condition, must be discovered and set forth by a singular Rod; which Rod is of two sticks held together with both hands: where there is such a breathing, it lies close on the Rod, no man is able to hold these sticks together, if that breathes on then: and if it were a single stick, it would break in two, the inner side where this Rod is laid together must be anointed with marcasite, that breathing draws it downward, even as the Magnet is of an attractive quality, to draw iron; so draws the breathing of this ore, the Marcasite; because the condition and property of the Marcasite is to strengthen the breathing of the purification. There is not a better way of Refining, as to use to each metal its peculiar Marcasite; there is a two-fold Marcasite, one above the *Uredines*, or upper elemental heat, and the other of the nether, of this marcasite *Schlich* or small dust is put to the point of the Rod the metals breathing causes the Rod to leap asunder. At melting of ores and metals there must be added a double slack, an upper and a nether, that the metal may break the better; the like must be observed here at this purifying and refining. The Marcasites, because (especially those above, not underground) carry soap metal, and are pure, help the purifying very much; pure ferment put to dow[9], ferments the same quickly; the same condition is in ores, a small addition will further their refining. There is a general complaint, that Marcasite requires a great deal of labor, before it can be brought to any good, what is the reason? The true manual is unknown. I call that a manual[10], when nature can be assisted with advantage and skill. Add to any metal or ore calcined, which you intend to refine, that is fit to be added, you will quickly see the effect. I could speak of the several Marcasites, of the several ores and metals, but it is needless for the After-works. I would not willingly have perverse men be acquainted with these secrets they have it often in their hands, but they do not know what they have; for what is Marcasite of iron?[11] Is it not the Magnet? And what is it

---

[9] This matches the 1671 edition. The R.A.M.S. edition has this as "dough." -pnw
[10] The 1671 edition has a side note here: "A Manual, what it is." -pnw

of gold? Is it not Lazul? And so forth of the rest. Go down into the mine, take this rule along with you, then you will come out of it more wise then you went down, and be thankful for it.

---

[11] Here the 1671 side note says, "The Marcasite of iron is the Magnet or Load-stone. The Marcasite of gold is the Lapis Lazuli." -pnw

Chapter 25: Of Furcilla, or the Striking Rod

As man's natural breath goes in and out, so is the halation also, and it carries all the rest. Man's breath when he drinks wine, smells not a little of it, or after any other hard scented food, this carries this breath in and out all other scenting breaths out of man's body; even so do I compare this halation also, with a natural quick breath, which carries all the rest, that come in the *Ferch* and seed. So there is no other striking rod for it, but what is of a year's growth, which otherwise is called a Sommer-lath, this the Miners cut off after the usual way, and so in the name of God they go: if it strikes in, then is it good; if not, then his unlucky hand and misfortune is blamed, which is fallen upon him, and thinks that his handy-works are not blessed; thus out of a false conceit men suppose their own aptness and disposition, either hinder or further them therein, not considering God's gifts and blessings; and the most of them do not know after what, and at what, those rods do strike, they wear it under their girdle, or on their hat-band, keep it in a devotion, as their superstition leads them unto, to get their bread according to their necessity; God has bestowed enough on them, if they knew how to manage their things judiciously.

There is one thing to be noted also, that the several airs or breathings have their several operations, especially upon aerial things. For astronomers appropriate some trees and fruits to the air. The upper air has the upper part of the trees and their fruits, the nether air possesses the root, milks and attracts the juice they yield, because it is their allotted part, as well as the upper part is allotted for the upper air, where it effects the juice, leaves, fruits, as it is seen, that the blessing comes from above at first, which is palpably seen in proper twigs, they make the twig and the fruit twistingly firm, and this sap sinks in due time, which is drawn up by degrees by a hot air from below, which is their food, their work, matter, and all aerial things.

Therefore, if you take such a rod, especially of a Hazel, or of a Kray tree, whose sap is full, and bears a pleasant and sweet fruit, it draws the same downward, that the rod must sink and strike, it will suck out the juice when the rod stands right on its passage, this holds from above unto this station, where the rod stands still, that place they call the ores station, standing as straight as a line.

This rod draws the halation naturally after the sap so strongly, that the rod does bow to the halation, and to that earth, unless it were so strongly pinned that it could not bow. It does not draw obliquely, and strikes neither at the upper nor the lower, but at the middle part, which is called the wind, and is good neither for the one, nor for the other, neither does it breathe, thrusting together the beams, making their outer rind, according as it is either good or bad, so it hurts or furthers the upper or lower breathing. This distinction must be well observed, the streams and ores run that way also. Hereunto are used the tops of Almond trees and of such fruit trees as do consist of a mere kernel, and a hard shell; for such shells, and the ashes of such trees, and all that they have, are much aerial and fiery, which serve best for the metalline work, and the best kindling coals are made of them, and are very light.

Chapter 26: Of Virgula Tripidans
Or of the Heaving or Trembling Rod

The breathings driving the fumes together above and below, you must conceive, that it requires great skill to discern other breathings from this; the breathings of the upper Elements are jovial, and the nether breathings are jovial also, the sane is apparent in their work. The wind also is *Jupiter*, he fumes these together, and that breathing can hardly be discerned by the rod: this skill is often a high strain; This heaving rod does show the way, it must be made of a metalline lamina, as of steel, which must be thrusted below the dam earth, at the end of that earth these breathings and halations come together, and the rod stirs, quakes, which it does not, unless at such a time. The rays of the inferior and superior bodies may not be stayed, but only at the work they do stand; for at their breathing they have their ascension and descension, are not held in their life's going, unless by the matter of this rod, there is a knot on that stick, and it is hollow like a Cane; this matter of the rod is made of an *Electrum*, drawn from the best thin layers of *Luna* and *Sol*, put into a glass head, which hinders not the breathing, because it will not be hindered, as a vessel may, made above ground; therefore look how you thrive, if you drive it too deep, if you take not matter out of the inferior earth matter, as has been mentioned before.

The strength you perceived of the striking rod, if there be an aerial halation at hand, you see how that rod quakes, for the inferior ascends and touches, and the superior descends and touches likewise, which rouses as it were the *Ferch*, or life and seed of the metal. The like appears in the clouds, how they move to and fro, before any snow or weather comes, and before the vapors be digested, which ascends: the same process have the halations subterranean, before they can mingle with the superior halations, they must be plucked off, emalged: if any of them ascend yet higher, they dissolve either into rain, or congeal into snow, dew, and hoar frost, before they part asunder.

Chapter 27: Of the Falling Rod, or of the Nether Rod

That there is a peculiar and vehement moving quality and power in the breathing and halation upward and downward, which may not be stayed by any opposition, the same is apparent in all vegetables of the world. Rain and great thick mists may keep from us the Solar shine, but nothing can stay this breathing, because it strikes directly through, but it proves weak in its fertility, it is the fault of the mild air; a shot out of a gun, if it is to go a great way through the air, is cooled. This is the reason why the Solar bears have not that operation under ground, as they have above it, and hence also is it that the metalline matter is not so fertile, however they do what they can, descend to the metal, where they rouse a fire and preserve the same. Thus, the Sun-beams must affect all in a magnificent and glorious manner not only above to preserve with its luster the lights of all the stars, but by day also to impart into the world and men the daylight. By these beams, it casts into the earth a dispersed fire, which is advantageous to the preservation of subterranean things, they thrust together homogeneal things, and bring metals to a compactness, that they may be had pure above ground, they rely and lean on them; from hence has the rod its original. For at some certain time you see the Sun draw water (as the vulgar speech is) so the beams have that quality, they comfort metals by striking downward. There is used a striking rod, above at its twist is left a part of the stick of three thumbs length, at the off-cutting, take out the pith, put some superfine *Sol* into the halation which goes in, pulls the rod in, where the ore is. The reason of it is, because the nether breathing of the air is there, and goes against, draws this breathing, carries it down, and the metal within the rod is pulled downward, as it would join the same with the rest of that body. Thus the breathings and halations are copulated; the beam must conceive on the earth, in which it is to operate, it brings no rain to the earth, but the vapors of the earth which rise, the same make or cause the rain, snow, and hoar-frost, bring the same to the earth, imbibing the same therewith; the same operation it has on the subterranean vapors, caused by the breathing of the metals within; in their ascension it makes them heavy, so that they are able to fall unto *Sol*, helping the same in its fertile work. Having accomplished its fertility of the work, then such metalline rain fall also on the metal, makes it ponderous, moves its *Ferch* in the *Lubricum*, and draws the rod down also more vehemently then it does to the striking rod which touches not the metal, but merely the breathing, which is not yet nobilitated through or by halation. This breathing might very well be called a fruitful rain, which thus comfortably refreshes them; it is not a running rain, but only a breathing which by the deep glittering *Sol* is made noble. It is not sufficiently expressed, when ignorant simple Miners usually say, that the Sun in her influences works gold not telling the manner of it (as I made mention of but now) themselves may go yet to school and learn their lesson better.

Chapter 28: Of the Superior Rod

As the planets have not their fixed and certain volution in the firmament, for the which they are called erratic stars, the lesser remain in their fixed stations; the metalline bodies are of the same quality; for they do not stick close unto bodies, but rise and fall ; for that reason I compare them to the various motions of the stars; above, their body is rolling and voluble, led and tinged by the *Ferch* and Seed, which is for the most part of the Suns condition. Planets with the Sun are of a special agreement *in motu & lumine*, and their body may be said to be *Lubricum & volatile*.

This Rod is appropriated unto planets which hang their bodies to the seeds as long as the great breathing of *Sol* holds still, and suffers himself to be obscured, and he must suffer in giving his luster to the earth by rain, mist, and snow, so long till he breaks thorough them. So, the bodies also are corruptible in this upper fire, more than the lower in their work. Therefore, such a rod must be applied which I call an upper rod, that the opposite breathing of the reflecting metalline beams which were received of the Sun, and strengthened again may be returned from the bottom to the *Systeria*, be it as high as it will.

This *Furco* or Rod, is a twisted tender branch of a Hazel, which from that twist on the bulk is hollow three fingers length, for its pith must be out of it, and the mercury of metals must be put in, the weight of three barley corns or grains: this Rod stirs the halation from below upward, this is the condition of this Rod, whereby is known that that metal is of an infirm body wanting the influence of upper planets, being defective of the Solar heat, and not of that below, and is the cause why *Mercuries-lubricum* in the rod is stirred, making it rebound; intimating, that it is unfit for his work, and that another body is fitter. The planets with their influence work more on imperfect metals, than on perfect ones. For the Sun does not put a garment on mercury of the body, but a certain constellation does it, the rest joining in the work, and are found next to it, except *Sol* and *Luna*, these meddle not with this rod. Though these also are planets and belong to a perfect breathing, and have their particular rod, as others have theirs, which press toward their several works, they are fitted for not aiming at the works matter, which makes its principles as the mercury of metals.

There must be had a sufficient instruction for to know it. It is thus; a metal is known before it casts any crust, and how deep the ore lies, and how it may be wrought upon, if I would know whether there be a sufficiency in that ore, the same I learn by the leap of this Rod; if it be a Saturnal ore, the Rod leaps the more, more than the condition of the body of Saturn is. For this rod does not stir the body, but that which is the most in that body, namely, the mercury of the body; for it yields good store of it, and may out easily another garment, and be a mercury of metals, and so forth.

If in that place be Tin and Wismuth, the rod applied leaps not so quick, but more slowly than it does where Saturn is; where a Copper or Iron ore is, there the rod moves more slowly yet; But this must be observed, that there is no breathing about metals, unless they be under ground; for when they are there, then are they in their works, whether they be increasing or decreasing, they cannot be without breathing then. Thus, much of the rods.

Chapter 29: Of Resting Vapors, or of their Sediments

Mine-works meet with many natural obstructions, as all other worldly things are subject unto. For all things sublunary are in subjection unto corruptibleness; the same happens also unto metals; they suffer by their cold or hot fires, that they quite evaporate and expire. As when their vapors and fumes grow ponderous, cannot vapor away upward: where such vapors are, Miners cannot brook underground.

In a cellar, where new wine, or new beer is on working, that vapor suffers no candle to burn, by reason of the ponderous Kb or fume, which cannot so soon enter the ground, unless these fumes be extenuated in their ponderosity. This fallen matter lies on close, is like warm meal, I call it *Caries*, which differs from an *Ostiocollum*, for that is in the rocks naturally, and dissolves at last into a fire. This does not so, and keeps still dry powder, and is heavier than mine-ashes, which are light, and dissolve into salts, which this metal does not, but abides a constant sediment, which is like an unwholesome stinking vapor choking the ores, turning them into rottenness being obstructed in their driving above and below.

These fumes are caused, when there are hard rocks to which fire must be applied, which can sethill[12] fumes, which being weakened in its rising its own in a place, grow ponderous, and increase; for evil and poisonous things will soon gather to a heap. It is seen in gold, being dissolved into water, let it stand a while, and the impure minerals separate from the essence of gold. Excrements of metal are soon brought to that posture; for being brought to a glowing by a small heat, they cast such a malignant fume, which choke men immediately.

---

[12] See pg. 68, 1671 edition. This is the exact spelling used. -pnw

Chapter 30: Of Weather-Salt, Halitus Melusus

The malignant breathing, as it is generally so called, is a pestilent salt, which endangers not only the Miners, but hurts the ores also; for it allays or makes the breathing fall, which are in the ore, being thereby obstructed in their natural working. The manner of the settling of these breathings are worthy of being noted. It may be compared unto an Eclipse, though the beam which ascends keeps its course, yet the breathing stays it, and will not let it go about that place; being in that stay and condition, its *Ferch* and seed part; at last, through its sides or walls it suffers fluid ores to come in; thence is the original of such fluid ores, of which there are several sorts. But this is to be noted; that weather is called a weather, because it is not of such clear air, as it is above ground with us, still carries something with it, which is grosser and hurtful unto man more than the air above with us; for men are not commanded to dwell and live undergrounds but if any ones employment lies that way, he must be contented to do so, commit his condition to God Almighty; because he must endure, and suck in ill fumes, and get mine diseases.

This subterranean air growing ponderous, by reason of the fume and water-stone, which stop and corrupt the cross passage, then it is called a Weather-salt: this is apparent when a burning candle is brought in, these ponderous fumes quench it like water, at last, they choke the men also. Therefore, when that is seen and felt, let Miners come out again, for they can do no good there.

Chapter 31: Of Cos Metallicus, or Stone Salt

The air, being of one piece of pure earth, at last turns to a stone. There are several sorts of Stones, as there are several bodies in the earth, though all of them are but an earth. The longer the matter lies, the harder and fatter it will be. Such stones cannot be stirred or heaved unless by fire, which seizes on the earth, and consumes it, and turns it to a terrene firmament, as scales are in fishes, and bones in beasts. These also the Lord will have in his good time with fire. This moved the Ancients to consume their carcasses with fire; though flesh soon comes to rottenness under ground, yet the bones do not consume so soon, the fire consumes them being turned to earth-ashes.

This salt is hurtful unto metals, taking from them their working, without it they cannot continue in their life, but must expire and die, because nothing can penetrate it. The difference between the stone-salt and the stone-meal is this: the salt heaves itself, when the stone grows to a hardness, which formerly has been a precious stone, and the air could penetrate it. But now it begins to turn into a horn-stone, when this meal comes to the height of its age, it turns to a dust.

Here it is requisite to relate the difference between the nether and upper stone, which partly is obstructive, and partly a furtherance to the Mine-work; they may increase under ground, but in the After-work no stone is good, because they have no nutriment, and therefore must perish and be famished for want of food.

Chapter 32: Of the Subterranean Pools

There is a most heavy case which befalls clefts, passages, and structures, called water-pools under-ground. For it is a truth undeniable, that clefts and passages under ground have their waters, both the day water and. the night-water, if once opened, then are they made passable, which prove the utter ruin of Mine-works. Therefore, my counsel is, not to make passages navigable to come to the ores. It is just as if one would dig a man's heart out of his body, and he would begin to dig at the pulse, and so a long all the way to the heart, would it not be a horrid butchery. There is a nearer path to the well, what needs to make far fetches about. There is a twofold water under ground; the day-water, and the ground-water; the day-water is not hurtful unto the structures, but rather helpful, soaking away many things, and takes away many times from the stages the ground-water; let the ground-water be kept out as much as possible may be, for it does hurt, if it cannot pass away, neither let it spring from below out of its *Cataracts*. Keep out the fire also; for both are hurtful to metals, and to all things in the world, if they play the masters.

Chapter 33: Of Aurum Metallicum
Of the Metalline Gold, or of the Metalline Bed

There is another fatness underground, where metals grow, neither is it the hull of the seed, nor the stone *Oleum Petroleum* or *Naphtha*, but is like unto a *Uredo* or Mine-soap. It differs much from soap; soap does not burn, because it contains a hidden sulphur, not a combustible one, but an incombustible one; this is the reason why it does not burn in a flame, neither does it seize on any other thing, but only on the pure metal, if that should not continue with it, the metal could not come, neither into ascension nor descension, it consumes first of all in the metal, even as grease does in the Animals. This fatness is not far from the ear[13]; when it is predominant it consumes the metal quite, and evaporates. Oil is of that virtue if put on the top of a drink, be it what it will, it keeps in its strength and coolness, this fatness shuts up the ores body that no breathing can pass. Fatness has a great affinity with iron, and is one of its next kind, of the which great volumes could be written.

In the County of *Schwartzburg*, at *Wackersberg*, there is store of such fatness, looks like quick-silver and swims and tinctures red, like Bole, this color it took hold on, and it may be extracted from it, some call it a mercurial body or a Tin-glass, lead-glass, Wismuth, or Antimony; it seizes on all those, and when this fatness does not turn to a fatness of such metals, where the seed is predominant, then it turns to a volatility, and to a great Robber. This fatness is a greasy earth, glittering like a salve of red and brown glowing, as if it were Quick-silver or beaten *Talk*, or glass strewed into. In the Rocks of *Bohemie* and *Transylviania* there is great store of it at *Goslar*, and at *Scalkenward*, even as quick-silver or lead-ores are; many of them are found here and there.

---

[13] Exactly as in the 1671 edition; probably a typesetting error for "oar", meaning ore. -pnw

Chapter 34: Of Metalline Streams

When *Ferch* and Seed must part from their work by reason of the sediments, be they what they will, and expire not naturally, then the ores turn to stones, which Miners call *Fluxes* though they know not from whence they came, nothing can be made out of them, they are fusible or fluid, when melted in fire, but nothing can be brought into them, because they are not porous, or no air can get into, which makes them more noble. It is strange in Nature if any good thing be driven out of the body, it will not return thither; for if life be gone from man, the body receives it no more; but these are things possible to God alone. My intent is not here to write of miraculous things, but only of things natural, I wave the former. It is to be admired that the body of dead metals is so fair, whereas other bodies which are dead consume away to nothing: metals also come into a corruption, but in a long time their death is like any glass, keeps its color, especially if it was of a *Marcasite*, hence are learned the colors of *Marcasites*, for green, blue, white fluxes are found therein, as metalline flowers have been, which are generated of three bodies.

Chapter 35: Of Creta, Chalk or Stone-meal

We see in this our air, that no fume or wind ascends in vain, it dissolves again into one thing or other, thither resort many meteors, the like meteors has the earth underground: For the fume which ascends from the fire halation of the ore, or of the metal, and affords the stone-meal (*Creta*) wherever it falls or lights, it grinds more, and increases abundantly having a dangerous salt, whereby it hurts those places where metals are, especially when they are in ascension, hindering their color. It is apparent in the flat at *Mansfield*, where it lies between the spokes of the ores, and can hardly be got from thence, it robs and consumes *Folium* and *Spolium*; the stone-meal makes a *Kuff* with stone-marrow, turning it to a kind of marble, called the *Petstone*, or *Doblit*, a double stone, and is dark and very firm, it strikes fire, being for the most part of fire.

Hither belongs the *Talks*, but intending to make mention of them in another place, I wave them here: however they also are such a meal, and differ from others herein, because it inclines more to a cold fire, wherein it melts like snow, as the others do, and dissolves sooner into water than into meal, and this turns sooner to meal than into water, if it be of less matter than it has of the stone marrow, then it affords a fair ice or crystal, called *Vitrum Alexandrinum*, or *Mary's Ice*, which cannot be mastered in hot fires, but it melts in cold fires, and is very hurtful unto metals; insomuch, that by reason of it Mine-works fall to ruin, as happened at *Stolberg*.

Chapter 36: Of Spiro, or of the Blast

The *Spiro*, or Blast is an Instrument which brings to right the weather or obstructed air, otherwise all would turn to stone, where it is, and would be at a stand there, if the lower fire should enter instead of the air, and exsiccate, though it does not kindle, if a piece be beaten off it, then it appears so, and this piece which flies thus aside gives to understand, how it makes the stone, and how nature frames the ore and metal: but jewels and precious stones are from another off-spring out of sweet waters.

In this instrument their dwell together fire and air, which take their power and matter from the malignant weather, where they consume all ponderous matters through fire, inlightning the remaining matter it has. Make that *Spiro* or Blast into a ball of copper, and an heads bigness, solder it bright and light, let no air get into it, leave a small hole, where a needle may enter, attracting the water, which purposely must be made and set for it: there must be had a pan of coals at hand, which must be kindled, and the ball laid into it, turning the little hole toward the coal-fire, and it will blow the fire forcibly; which being done, it grows hot, and makes the water boil in the ball, which fumes and carries it forth with a great fierceness, blowing on the coals strongly, and thus it maintains the fire by breathing strongly in the manner of a pair of bellows, driven from without: hereby several good things are effected, and, the condition of the ball is, that it shows what may be done above ground with the like, no use can be made of it behind that place, because nature herself has such a blast for her fire.

Chapter 37: Of Pulsa
Or of the Break-stuff, or Brittle Matter

This salt is engendered usually by malignant fume, which the mine-fire should have; and when the stones be very hard, then there must be made a fire of wood, where the fume draws to the stone-fire, and grows thick, and if the fumes of *Succinum*, and of other things are joined, it turns them to such poison, that the ore must be aided, else it perishes, for that fume lies on the ore fuliginous *Kobolt* which corrode and consume the ore; there a ball must be applied, which is round and hollow, having a hole at a bigness at which a quill may enter, it must be so close that no air may either enter or get out, this ball must be filled with gun powder, cover the same with a cotton-wool boiled in Salt-peter, then dip it in melted pitch, which is mixed with some Sulphur, kindle that ball, let it go down in a box, or fling it on a *Stolln* or chamber, when the ball flies asunder, it expels that fume, not only by that smoke, but with the blow or report the gun powder makes. Such a ball may be applied also to water, and be sunk in an instrument underwater, in which noisome fishes are feared, its crack will kill the fishes that are there; there is a paste which gives no report but only burns, and destroys, and heaves this Salt, but have a care what paste you make use of, and have a respect to the upper-scaffolds, whether they be old or new, that they be not embezzled, and your paste must be mixed, so that it nay do no hurt.

Chapter 38: Of Clathrum, or of Blank-fire

This fire needs nothing for its food, shines in darkness, is a special fire for Mine-works, quits the charges, if applied, more than the expenses do, bestowed upon Talk or *Bromith* work, for oil in some places nay be had cheap enough, casts no smoke, destroys fume: it is put into a glass ball, which is put into a basket, to keep it safe from water and sand, which affords a light to the work-men. Miners ought to know how to enter their ground for the metalline *Speculum*, which is a singular manual, for the metalline breathings, and after-halations join and come together; require special instruments whereby they may be known; for where the do join, and the diurnal breathing is predominant, then it exhales by day, shining out of the Earth; Miners call this a metalline breathing: True it is so, but they leave out something, so it is but half a breathing; if the after breathing is predominant, then it appears by that *Speculum* and light wherein it makes itself known. She is in work with something, and there is at hand such a metal, metals do shine, though it does not appear so to our eyes, like as rotten wood does, by day they are not quiet, as long they are working, but there must be reflection of their work, which is this light. It casts no beams as the day-light or rotten-wood does, by night it receives one from such a dark or duskish shadowed light.

Fair and curious breathings are seen therein, and that light of darkness is a light you may see by it, he that is distant from it five or six yards sees it not, nor can you either, for it is such a light, as is in the eyes of Cats, Dogs, and Wolves, which can spy you, though you cannot see them, for there is a light at night as well as by day, which is apparent in these bodies, which receive their light from the Nocturnal night, for if that power were in themselves they would ejaculate beams, which they do not, and experience evidences it, that there is a subterranean *ignis dispersus*, a scattered fire.

This light is twofold, the first light, being thus prepared in a ball of some fishes or worms, of juices of herbs, and saps of wood, being distilled, and the distilled water being put into it. Take a pure Crystalline glass, it casts a curious light under-ground, if mercurial water be put into it, it graduates the waters made of worms and of woods very highly in this darkness, which is called the light, it may be done and used also by day, but much better in subterranean darkness, in which the fire lies hid, and must be roused and awakened by such material and instrumental fires.

The second light is *Speculum*, which receives that light, and gives an intimation of such hot or cold fires, which not every Clown or Miner understands: For as it shines in the *Speculum*; so kindles this fire, and is the ore. In man's body are they discerned well enough, from whence the diseases have their several names, but are not searched into. The difference between the ball and the *Speculum* is the same with that which is above ground, I can view all the members of my body, but not my face, I can behold the light, but what the Sun of this light is, which ministers the luster unto it, the same I cannot behold or discern.

Chapter 39: Of the Gluten, or Mine-glue

The best help and remedy, which may be applied to subterranean pools, are wells: for where these break forth, they carry that water away: a better and nearer is not than the Gluten, to dam up or keep out the day water, that they do not run any farther; this damming has great utility; it makes the water not only slimy and rough, but drives it backward, that it be served for some other issue, and be rid of it in that place, where it is naught, and merely obstructive.

If the day-water be thus stopped by a Gluten, that it cannot run and gather at the sink, then that ground-water may soon be drawn away at the sink, the deeper the sinks are cleaned, the more these ground-waters or springs are diverted, and at last are turned also to day-waters, or may be dammed up, and made run another way where they may not be obstructive to the scaffolds, and where drivings of mils are not had at the same places, the Gluten may be used, then the Scaffolds and Structures in the passages, clefts, and Mines may be seen, the dams and Gluten are the belt helps hereunto.

Chapter 40: Of Truta
Or of a past for to Corrode the Stone through, or through Eating

There is almost nothing which is a greater hindrance unto Mine-works, than water is, and where the gluten is not sufficient to keep it out, and in places where it is shut in, and must be drawn away with lower buildings, as with *Stoll*, or beams and pipes: it is a huge and dangerous work, to make these thorough-breathings good and holding; it costs many men's lives, and great expenses must be made, therefore ways and means must be thought upon to make ways through with burning, to make such a fire which corrodes the rock, and grinds the stone, eating it small and thin, that the water may get thru, and run away, that the Miners may not lose their lives in that water, as usually it befalls them at such works. This fire corrodes great stones in running waters or rivers; it is a corroding fire, a Gluten being made, which is lined or covered wi.th combustibles, poured or cast down through a channel or pipe, guarded from water, that Gluten may be effectual, though it be under water many fathoms deep, it still corrodes further, gathering strength by that, it catches upon and burns, and presses still lower, it does not smoke, being a running corrosive fire.

There are some saps and gums, which if boiled to a hardness, and mingled with unslaked lime, kindles and burns so strongly, that they corrode the rock, make an hole into, as big and as deep as you will have it, so that the water must sink away, there must be set a pipe of wood or of other materials, as deep as the water rose, and must be set and sunk to the very bottom, and of this Gluten, Paste or stuff must be put into, let the hole of the Pipe be closed with pitch, to keep out the water, make small bullets of this Paste, kindle them, it eats down even out at the *Stoll*, or beams end, the bigness of the hole must be according to the Pipes mouth below, which must be equally wide with that above, when the Pipe is cleared, and way made for the water to run out at the hole, then all that water-pool under-ground will sink away, and clear the chambers below. This is a curious skill for to break through rocks, if well contrived and well-ordered with exact manuals.

Chapter 41: Of the Traha
Heaving Materials used instead of a Dray or Sled

It is known, that breathing, and halation, and the weather uphold all both artificial and natural things: it is apparent in great Edifices, that the things exposed to weather cannot hold, if neither water nor wind tied, the great reparations in such structures signify so much.

There is a place in *Zips* or *Sepusium*, called the *Tobeshaw*, where firm steel ascends by day, and in that place there is no Mine of steel, no instrument can get any scale from it, but lying in the weather one winter and summer, it gets a scale of two fingers thick. Thus, it is apparent, that the weather heaves also a *Stoll*, or the great beam or metalline body, why should it not lift and heave a stone. This appears further at a falling down of great snow-balls from mountains about *Saltzburg*, and in *Styria*, where great pieces of rocks fall down with such snow-balls, as big as a house is, which heat and cold has thus corroded and loosened. *Hannibal* making the *Alps* passable for his army, poured warmed vinegar on the Rocks, whereby he made them so brittle, that they soon could be wrought thorough; oil does the like, if well prepared. *Acetum's* made of vegetables of wine, beer, fruits, are precious for such purposes. *Cistern* waters may be turned into *Acetum's*, if cocted with honey, being made warm first; this drives the fire back, which is in stones, for there are commonly *Horn-stones* and *Fire-stones*, which are made brittle by such means. There is made a *Petroleum* also, so you need no sallet-oil, nor any other; no, nor *Naphtha* either, drawn from *Osteinmark*, or calcined flints, such water-*acetum's* being poured upon, and other frightening waters, whereby the hardened flints are terrified and made brittle. It stands upon natural reason, that such stones must be dealt withal in this manner; For behold the *Gluten* and Aquafort, of what efficacy these are? Does it not corrode the *Pumice-stone* like Bees-wax, and the *Top*-stone like a marble of diverse colors. Consider well the Whit-marble and the sliding sand, in which the Pumice is, you will find what manner of *Lixivium's* may be boiled from them.

Chapter 42: Of the Frost in Mine-works

The greatest troubles that Miners are put unto is to pull and draw up the filths and stones that are naught, out of Mine-works, that a way be made to come deeper in. Above ground they call it a heap of rubbish. It costs no great matter to cleanse, dissolve, and void these rubbishes with corrosive waters; it costs little, if rightly managed and handled, to dissolve first the lightest things, these being made riddance of, the rest may easily be voided. That earth under ground must not be looked upon as that is above with us, adorned with grass, for under ground there is least of the earth, there is a mixture of all manner of things, as salts, juices, minerals, stones, the least part is earth, and yet that part is the noblest, for out of it are made all manner of metalline bodies: There are sharper things, all of which must not be used at once; and must be effected with these, when that which is above cannot be applied to that which is beneath; juices also are easy in their uses for to corrode and make brittle. Sulphur alone performs the work, which is a poison unto juices and saps. Miners and such that are employed about such works must have knowledge of such things, and exercise themselves herein by way of practice; for all particulars belonging to these manuals cannot be set down upon white and black: experimental knowledge must be joined hereunto, not only a depending from things written.

Chapter 43: Of the Flaming Fire

Whereas there is occasion for great and small fires in Mine-works, which must be learned and applied accordingly, to the several sorts of metals, and not after the manner of their several meltings and firings, and the condition of such necessary fires must be known also. To set down these in their particulars would require great pains, and the writing thereof would rise to a great volume: it is the duty of understanding Melters and Finers to order and regulate themselves in their fires, according as each metalline condition requires, to further and not to hinder their work; and so, I commit these to their further and serious thoughts, and to take these things into a fuller consideration.

Chapter 44: Of Ignis Torrens
Of the Roasting Fire

Things inclining to ashes, and soot, and excrements of metals, anal the *exuviums* or hulls of bodies Melters suppose nay be taken and gotten off safely in a roasting or calcining fire, they make a great fire of wood under them, roast, or calcine the metal , that as they suppose they retain nothing thereof, or of such offal you heard of now, they yield their *exuvium*, and copper yields cinders and slacks; but if frightened, then it robs and consumes iron; therefore nealing is more commendable, as they do at *Mansfield*, a great heap of ore is laid together, which they kindle, let it stand in a gentle glowing heat, and burn away that which should come off in that glowing. Metals in *Swedland* are healed thus at the heat of the Sun in Summer, there it runs finely together, and purges itself so neatly, insomuch that it would be refined, if it staid its time in that heat. This nealing I do better approve of than of the calcining in a fire flame. There is a twofold glowing fire, and metals require a twofold glowing or nealing, one sort of it is used at *Mansfield*, they kindle with bundles of straw the heaps of flats, let them glow of their own accord, and they do it like a heap of coals, and the ore is nealed, which is put in for that purpose. Secondly, nealing is good also for bodies of stones, reducing them into calxes, but those that made metalline calxes in an enclosing heat, or glowing fire, they got only the calxes of the bodies *exuviums*. Therefore, neither themselves, nor others have any cause to marvel, if they do no good in that way.

Chapter 45: Of the Corroding Fire

This fire ought to be set among the coal fires, being of a consuming nature, and their corrosiveness is in the cold fire, and it has the same qualities which the burning fire has; it shines and burns; its burning is corroding, in that it is better than the other, because it does not burn it to ashes, but bring the bodies to a dust or sand, which would be toilsome, if by filings it should be brought to stars; the next neighbor to this fire is the glowing fire, of the which I will give only an hint.

Chapter 46: Of Ignis Candens, or of the Glowing Fire

This fire is purposely ordered upon metalline bodies, it consumes them, being their matter is naturally inclined thereunto: This fire is of great concernment, making their bodies very malleable, their *exuviums* stay on the Float, and is the best quality they have, that they put off in that glowing the thing which will be gone, and the good thereof remains. Things nowadays are slighted, the world supposes to have skill enough it wants no further knowledge, *Quot capita tot seinsus*; everyone thinks his wit best, though some have scarce begun to know any of these things; which is the reason why men are still kept to their rudeness. Men may suppose, I mean by this corrosive water an *Aquafort*; it is no such matter; how many tons of precious *Aquafort* is used in vain at *Goslar* on the *Hartz*, which would serve for better use, and the expenses laid out for wood might have been saved.

Chapter 47: Of Ignis Incubans, or of the Lamp-Fire

This fire serves when metals are wrought openly, and not luted in, then the metal does not fly away in a dust, nor does its best run away; for you heard that a flaming fire is hurtful, for to work metals withal. Lamp-dishes are commonly of glass, set in an earthen pan, filled with ashes or sand, kept in a sweating, in that sweat many suppose the metal receives its body, or the one changes into the other: I leave this transmutation in its worth, and cannot approve of it. Touching this warmth, I cannot disprove nor find fault with it, and all metals indeed should be dealt withal in this manner.

These two fires of ores and the Lamp-fire, if they were made use of in Medicinal ways, would do better than calcining or flaming fire can do, where these are of no use, and the long fire must orderly be kept in an equal heat, if any good shall be done. Some kept the Lamp-fire in a Stove-furnace, where all things were spoiled in the working, it was either too hot or too cold. It was of no equal heat, which the work in the end did show, because it was not well governed.

Chapter 48: Of the Cold Fire

This is a strange fire, little can be said of it to those which cannot conceive of it, whether it was not taken notice of, or whether they did despair of it I know not: this is it which elsewhere is called coagulating; it cannot consume the other fire; it can melt the work, but to consume it is impossible, it works in the air as well as in the fire, where it shows its efficacy and is the sole proof of its subtleness; metalline mercury is of a cold fusion, all other fusions are hot; if you believe it not, feel it; the fixation of the warm flux is called coagulation, there the one opposes the other; the one congeals, the other keeps in a liquidness: this difference must be known by those which are employed about melting of metals, and their fluxes.

It is of concernment, to govern this fire well, or how stones are to be weighed, and things that are excessively cold are a death to a tempered body: what animals do live either in too cold or too hot a fire, and to speak precisely of life, it is impossible to do that, as to speak really of God. Therefore, gaze not upon definitions, what humane reason is able to conceive of: Philosophy is strangely conditioned, and it appears by this fire also, a thing which is very cold, may contain a life however.

When it is in its highest degree of ascension, then it comes down again, it turns to silver, then to copper, if the nether hot fire does it not, then surely the cold fire must do it, for it dissolves again into its mercury, which is the flux of the cold fire, if it lays hold on it, then it must run to all bodies, in its running it puts off not only the nethermost but the uppermost body also, take this into further consideration.

Chapter 49: Of the Warm Fire

Of this I have spoken already, it can be made and governed several ways, coals, wood, pitch, oil, and other combustible things are fit for it. There I would only speak of a heat, which is good for the flux of metals, whereby they are purged, as you heard above: needless to be repeated here.

Thus, much of this first Part, where I informed about the nether work, or fore-work, governed and observed by nature, whereby she holds forth unto us metals and minerals in their forms. He that conceives aright of this work and considers it, works with advantage and utility, and is a great help to proceed successfully in Alchemy, which imitates and treats into her steps. I wish hearty success to all such, which bear an affection, and love thereunto.

Praise, Honor, and Glory be unto the Supreme Master of Mines,
By whose word and will all things are made, ordained,
And brought to their forms, Amen.

End of the First Part.

The
Second Part
of the Last Testament
of
Basilius Valentinus

Friar of the Order of St. Benedictinus.

Wherein are repeated briefly some principle
Heads of the first Part, what course
Nature observes under ground, and how
Metals are generated and produced to
Light; as Gold, Silver, Copper,
Iron, Tin, Lead,
Quicksilver,
And Minerals.

In like manner of precious Stones, and of
Tinctures of Metals, how they are discerned,
And what relation they have to the
Holy Scriptures.

LONDON,
Printed for Edward Brewster, at the Crane in
St. Pauls Church-yard, 1670.

The Second Part
OF
LAST WILL AND TESTAMENT

Chapter 1: Of Mines and Cliffs
And what Manner of Middle and Second Works are in Ores

In the first place there lies a necessity upon every Miner, to know how to search and dive into metalline passages, how they strike along, and they must be well acquainted with all their occasions and conditions; and if at any place he intends to fall to work, he must know how to use the Magnet of the Compass, where East, South, West, and North lie, and learn the ways of this and that ore, and where their issue is, and be well informed of the long and short strokes of metalline passages, where they draw together to a metalline form. The forms of metalline ores are several; some carry Talk flats, an ore which contains silver and lead, others are very brittle, having little of slate and talk, and these as discerned by their firmness; there are other stones in which appear Copper, and the flowers of *Zwitter*; there are others also which have flat *floats* and Slate-stone, in which is wrought Copper ore: hence it may be gathered that by reason of these several forms, are produced several fruits; and in Mines toward the South better ores are found, then there are some toward the West called after-ores; between which there is ordered or placed a center of perfection.

Chapter 2: Of General Operations of Several Metals

The Almighty for his eternal honor and glory has held forth to mankind innumerable wondrous works, which he as the sole Mediator and Creator has set forth in natural things, the same he has showed also in his omnipotency underground, in the metals and minerals, of them we may learn, as the twelve Sybils prophesied of the bright, true, and only Son of Righteousness and Truth, in which do rest after the twelve ports and gates of Heaven, and after the twelve months, moveable and unmovable, visible and invisible bodies, the seven Arch-Angels standing before the throne of God; after these the seven planets, Sun, Moon, Mars, Jupiter, Venus, Mercury, Saturn, and the rest of the stars, and the seven metalline ores in their properties, as Gold, Silver, Copper, Iron, Tin, Lead, Mercury, then Vitriol, Antimony, Sulphur, Wismuth, *Kobolt*, or Brass-ore, Allome, Salt, and other mineral growths.

That the true center may be comprehended and conceived of, God has made the first separation according to his word: the Spirit of the Lord moved upon the water, the whole elemental body of the earth has been water, but the Spirit of the Lord *Zeboath* has divided it, and fashioned the earth from the muddiness of the water, and therein all metalline fruits that ever were created and generated underground, all these were first water, and may again be reduced again into water; all other creatures, be they animals, vegetables, minerals, all these are produced from the first water, the several kinds of beasts, fishes, and sea-monsters, after the Lords Spirit, and after the first eternal breathing Essence, which brought forth and shaped things tinged and untinged, soft and hard, small and great creatures; after the twelve stones in the breast-plate of *Aaron*. He created man after his own image, the holy Spirit was infused into *Adam*, who had a fullness of eternal wisdom, and that according to the order of *Melchizedick*. Almighty God, who is the first and last, the first principle and end of all things, has set his gifts into times and hours, days and years, which according to his eternal Decree have their revolutions, he has blest in his most holy means *Abraham, Isaac, Jacob, Aaron, Melchizedick*, and others he has infinitely blessed, according to his good will and pleasure from eternity, putting several periods unto them; and in his unsearchable decree and will he has laid the foundations also for Minerals and Metals, a help for the supporting men in their necessities in this miserable life; thus has he meliorated and exalted the earth in her goodness, men have reason to return hearty thanks unto the Creator for it.

God in his gracious providence, next unto the knowledge of himself, and of his holy Word, can bestow no better gift to man, than to imbue him with the true knowledge of Metals and Minerals; Jews thought themselves wise men herein; but as little some Miners know Minerals and Metals, as little knew the Jews their Messiahs and Gods word in its true sense. Therefore, from that blessed and promised Country the knowledge of precious stones, minerals, and metals are come to us, as by an inheritance, as being the last, and are become the first, and they the last; but in the end Heaven's gate will be opened unto them again, internal and external gifts and means will be bestowed on them, and the true use of metals will be none of the meanest.

Where there are fertile stones, be they rocks, flints, pebbles, marbles, in their central points is found what they are in their operations. The several gums and resins, the one excelling the other in beauty, transparency, hardness, or liquidness, are known and discerned by their fragrancy and salt: Miners ought to endeavor incessantly, and in simplicity, how the nearest way may be chosen to find out the Mineral-passages and veins, into which God and Nature has laid direct courses.

Chapter 3: Of the Stones, Rocks, and Flints of Gold
Its Operation, Condition, and Striking Courses

Gold is wrought in its proper rocks and marbles, and in the purest matrix of the firmest earth, of a most perfect salt, Sulphur, and Mercury, purged from all *feces*, and impure spirits, with the conjunction of a natural highly clarified Heaven, of white, yellow, and red sulphureous earth, after the fiery nature of *Sol*, in a deep fixation; insomuch that none of all other metals has an higher, compacter, and more ponderous body of a goldish matter, in which there is no humidity; all the elements are equally in it bound up, which in their unity have wrought such a fixed body, tinged the same throughout with an everlasting citron color, with the deepest tie and uniting of its pure earth, Sulphur and Mercury; and with its Vitriol essence it does all, what the Sun among the Stars does operate. Naturally, all is gold, what cleaves there unto in and at all sides; and it is found in the best and closest stones and passages, and the power of *Sol* works merely upon that ore, and in its quality, is comparable unto *Sol*. This noble gold stone and ore is sometimes mixed, and on its outside there sticks some obscure and dark matter, having annexed to it some flats and other spermatick matter, which detracts from the goodness of its own nature; and though the Creator has imbued it with great Virtues, yet it humbles itself, and suffers itself to be found in despicable Mineral-stones, where it loses much of its tincture, as is apparent by the touch-stone, where the mixture of Copper, Silver, Tin, and others are seen; all these mixed impurities can be separated from it with artificial manuals, and with little ado it may be brought into a perfect state. Gold ores naturally are wrought thus, that the gold stands in it close, compact, firm, and good, which is found sometimes in the cross passages. Its fixedness is found in the deepness underground, where it has its greatest power, and it is found also sometimes in a speckled jaspis, full of eyes, and mixed with flints in its passages, where many times vitriol flint is found abundantly; which Vitriol is the best among all other sorts of Vitriols. The Hungarian Vitriol has the precedence before all the rest, which is sufficiently known in their proves and examines, as may be demonstrated to the eye.

In its passages are found sometimes fluxes of several colors, which are interlined with gold, and must be forced with fire. To that purpose, it is requisite that it be dealt withal with such fire, as you heard in the first part, commonly *Zwitters* and *Zirn* stones are such, which must be stamped and beaten, and drawn to a narrowness and fined.

Gold is wrought also in standing passages, and on level ground, the ores and such passages are yellowish, rocky, and of an iron shot-sand in cliffs is it on-grown compactly, and generally it is found near Flint-works, sometimes it is found in a flint, or in a liver-colored jaspis, sometimes in white pebbles, that gold which is in it is of a white color, like silver, or white-copper ore, where it sticks hoary and rugged: it is found also in brittle Lime-stones, where it stands curled with black specks unsprinkled, is granulated, like drops found in the subtlest firm stones, spotted with iron moals or spots, and are protruded in fair yellow flowers, and are a black exhalation thrust out. It is found also in streaked flat-works in pure passages, mixed with a blue Horn-stone and flat; in flinty glittering passages it is found hoary and compact wrought. There are found also flat marble floats, wherein in all your cliffs is wrought inherent gold, mixed with green grit, and iron spots; sometimes it is found also in square iron shots, or porous marble Marcasites; but for the most part in grits, sometimes gold ore is found also compact and firm in black passages: some gold ores and gold passages are found also to be of Minerals and of Vitriol, and Miners in *Hungary* especially can discourse of it, because gold ore is found in that manner in those parts.

Chapter 4: Of Silver Ore
Of its Mine, Operation, Conditions, and Striking Passages

Silver ore is wrought in its own stone, of a perfect nature and most noble earth, and of a fixed clear Sulphur, Salt, and Mercury, which with a mixture does join in a fixed and firm uniting, and appears of a degree lower than the gold is, and is the best metal next to gold, and in the fining of it, it loses very little, and is separated *per se*, or with other metals joined in the fire, its natural fitted stone causes the silver ore, following its heavenly influence, and the nocturnal influences of the Moon. In Northern parts the most silver passages are found; for as *Luna* borrows her light from *Sol*, even so the silver passages and silver stones, have at their right-side Gold passages, and with that noble Queen *Lunaria* is compared a root, whereby the gold passages acquire strength, and get the more power in that mixture, and get their ores from their roots. Ancient Philosophers wrote strangely of Virtues, to be a fertile yoke-fellow of *Sol*, which may be applied to the upper and nether metalline work, because nothing is so fix, next gold, then silver is in its perfection, and is the reason why silver passages are accompanied with white fluxes and mineral veins, next thereunto are such passages, in which are generated red Mineral-Sulphur, and red yellow juices of the noble gold.

Silver metalline ore is wrought many times in a red goldishness, and comes forth better than the other, a proof whereof may be had, if well ordered. White gold ore is naturally thus tinged of white copper glass, which cause such ores and passages, by reason of the food of their perfect Minerals, and with the glass ore black fumes are exhaled, and feed. upon Wismuth, Lead, and Tin ore, wherein Minerals that strike near upon the Lunar passages, are greedily refreshed; thus grow the firmest and compacted silver ore of its pure, proper, and unmixed stone, meliorating the bad places and instruments which silver ore has many remarkable virtues next unto gold, from the heavenly influences, changing several sorts of silver stones, descending from the originals of their highest finished unity.

They carry and produce also, not only mixed chambers and Mine chists, but also several hard and sturdy mixed ores in whole flint-works, and other copper-flowers, yellow and black ore, and are found different in their nature, form, and tincture, so that the one is more hard, sturdy, slatty, broader, narrower, whiter, bluer, in its color thus qualified and natural in its end, middle, and beginning. This is the reason why these silver fruits and ores are found differing in their colors and forms, the one being more compact, fairer, and of a better glass, than the other. Sometimes there is found in such a vein or passage firm and compact gold, silver, and copper, so it is found sometimes at *Krenach*.

There are found and seen also in a certain vein and passage in mixed Lime-stones, lead, iron, and copper ore in union and juncture. And in one mine is found copper ore, in another is found silver ore, and in another mine is found an iron stone; why should not such remarkable distinctions be taken notice of, which Nature from Gods imagination has held forth so gloriously unto Miners, and set these before them to be discerned by them. Some silver passages are found also in their natural *Zachstones*, which either are in the hanging or lying ones. Silver passages show themselves also with blue gritty flowers, hollowed fluxes, in sprinkled Marbles, and carry flint-works of several colors, and these passages and cliffs are full of pleasant silver colors, of yellow and green, of a color of *Goslings*, the more they are mingled with such colors, the more they have wrought.

There are some silver passages and veins, which carry three distinct colors after the manner of a rain-bow, where the one-color works in nature either more closely, or more mildly than the other, in a curious order, and the one may be discerned before the other in their passing streaks and shootings, together with their Chamber-colors and floats, as they fall severally and apart in each mine ore.

Chapter 5: Of Copper Ore
Of its Stone, Operation, and Striking Passages

Copper ore is wrought in its own and proper stone, of good pure salt and over-hot burning Sulphur, through a heavenly impression into all its parts, tinged red throughout, not quite freed from a superfluous humidity, in an affinity with iron, because copper and iron are nigh kin one to another, because their dwellings and houses are set one by another, and is the reason why the one may easily be transmuted into the other.

This metalline ore is much wrought in flat float-works, which are green flinty; many times it appears in a red or brown form, and is seen also like lime-stone in black and yellow flat-work; like unto coals in green flinty passages, in a twofold manner, either current, or in a manner of a float: sometimes it is red and brown, mixed with a green color, some are of a lazure color, some of a copper glass, flinty and iron shot, or of a white food. The copper ore in its passage is sometimes rich of gold, and of silver, and is accompanied with curious *Zach* stones, and enclosed with passable stones, if so be that other metals arid minerals do not entrench upon them, which corrode and consume them. And copper-ore is a flat-work also, mixed with foliated earth, and the mercurial copper is hardly brought out of it at or in an ordinary melting, affords store of iron, and unripe copper-food, which rob very much the copper in roasting, and make it unmalleable: The richest copper-ores are found in *Hungary, Bohemia, Silesia, Thruringia, Hassia,* and *Voigtlandia;* the like is found also about *Trantenau,* where it breaks everywhere in a manner of a float, mixed with sand ore, and where it breaks vehemently in the flat work, they call that flat of cliffs, they are poor in silver, and such must be roasted or calcined, in some places it breaks in a fair blue and brown color, or it looks ruddy, of a copper glass, and like unto green oaker, and sometimes it is white goldish, which is called white copper-ore; it grows white at an effectual mixture, because at its uniting it assumes or takes in much of silver and of lead: it breaks also of a yellowish and lazure-like color, green flinted upon floats and moving passages, in lime and spongeous stones. It breaks also of a blue color blue oaker, is copper, glassy, and flinty, in great and huge rocky and marble passages, being mixed with a white marble: they are rich in silver, in green flat stones which are clear and brittle, it lies dry and green in cliffs, open caves and passages, like green frogs insprinkled one in another, in a strange manner, distinct, or parted with strange pleasant colors; which graduated works are losers in half their worth; in these rocks are strange cliffs of marble, and of white veins, yellow flint is insprinkled and mixed with copper passages, which yield much silver, have few flowers, are of a ponderous form, break very flinty, of a red glass, of a green color nixed with yellow flowers, these flints are joined with white gold marble, of a green color, besides the rocky passage.

There is found also copper ore which is rich of silver, flinty, and not white goldish, is of a white shining glass, mighty in dry hollow flat mines, some whereof are nixed with iron, or sorts of Wismuth or Fire-stones. At the one hanging of some passages is wrought the *Chrysocolla* and copper ore; on the other hanging of the mine is wrought pure flint, all according to the quality and condition of the ore. And it is to be observed, seeing that copper ores are usually mixed with Sulphur, easily unite with the nether metal, and join with their stones, therefore green flinty copper ore which carry in the dry lead, slatty passages, a black *Molben* are Mineralish, and are not rich in silver, nor rich in *species*, encompassed with immature iron and perfect copper ore, and some are free of it, if far separated asunder, from dry mineral flats, are richer in gold and silver, according as the stones take, in a good natured ore, they usually entrench upon gold arid lead rocks, or antimonial ore, as also upon iron and silver stones. There are found also flinty passages, that have their Mineral juices or Vitriol, and Sulphur; some whereof partake of allome *& alumen plumosum.*

Those commonly have the best and most copper passages, which are least mingled with other metals, as lime, and tartareous stones, in which black floats and flats do break, are enclosed with green, and are of mild quality, at *Eisleben* and *Mansfield* Miners put their several proper names to it very exactly, according unto their nature. Miners in *Misnia* know least how to distinguish these, the upper part of clay-earth they call *Putredo*, in which the true earth is also, and when they come to the stones, they call it the *Day-work*, because they cover all the rest, and turn quite to stone. The third place they come unto they call *Night-work*, because it is easily lifted and heaved one after another, and is pure, then they come to the Cave or *Hole-work*, which must be hollowed and set, here are the stones which must be broken, then they come unto the flat, and below that flat they come unto the sand ore, though sometimes it be on-grown at the *Lochwerg*, or hole-work above the flat, then they turn unto the dead earth again. Slat and richest copper ore at the silver breathing, lies also on the rocky, horn-stony combustible ores, which have their gold and silver passages of your special kind, among which there are found several forms how each of them is discernable. In *Hungary* and *Carinthia,* the passages yield copper ores, which copper is very malleable, and is at a dearer rate than any is in all *Europe*, as their Minerals also, and especially the Vitriol there is held to be the best: as also their Antimony is counted the best. That vitriol has the best and rarest virtues, which is known to true naturalists, and experience has proved the same to be true. I speak something now, which if reason and Understanding were answerable, many expenses, hard work, and good time could be saved, and it comes only from hence, because Gold breaks so near to it, at the same ores is found, where that earth is impregnated with goldish seed, and make use of the same food in many subtle unitings. Minerals in their generating qualities are better supported among perfect metals, where they are higher and more effectual, and are best used for both such perfect metals, in case Nature be rightly imitated, the ancient Philosophers have had experience of, and made trials of it. There is a remarkable difference found among Minerals, and partly from copper ores; they are Minerals and Metals, each their particular nature and being, among which some ores look green, and bleach at the day, and grow near other metals; but their stones are most like unto lead-stones, some whereof are grosser, softer and harder than others, and some are more obscure, dark, muddy, and some are green, and so forth.

Chapter 6: Of Iron-ore
Its Mine, Operation, Stocks, Floats, and Passages

Iron-stone and iron-ore is wrought in its Mine-stone, according to the heavenly influence of *Mars*; for he is *Trinus magnus*, the great Lord of war, and an instrument whereby others are forced and compelled; of a hard, earthly, impure sulphur of putrefied salt and gross Mercury, which three principal pieces in their juncture mix much of earthliness, therefore is it a difficult labor to mollify iron with or in the fire, carrying much of impurity by reason of its sulphur, and above other metals it has a deep red quick spirit, which if it be taken from *Mars*, then is the iron gone also, leaving again a putrid earthliness. Iron is not easily mixed or joined with other metals, or united in their casting. Iron has a threefold partition, and several parts in its earthy ore, namely, a Magnet, a quick metalline ore, which has its quality from quick Mercury, and must hold communion affinity with iron, must be quickened and renewed with iron filings, in which he lies like a hedge-hog, and is induced by the Sun of Nature with glorious gifts and Adamantine virtues; at one place and side it attracts, and at the other side it repels, which virtues may be augmented and increased in it; it plainly typifies or demonstrates (like unto the Sun in the heavens) the true hour in the body of the Compass, by water and by land.

Secondly, steel, the hardest and purest most malleable iron, of its proper light draining place, wherein it lies close, tied and knit together, in all its parts most compactly, which in all iron works is usually put to the edge and point.

Thirdly, there comes the common iron-ore, ordered together by its earthly Sulphur, which three ministered good thoughts to the first expert Naturalists; that Master of Mine-works *Tubal Cain*, who made his three principles in all things, and made his dimensions in the mines in three distinct parts, in which such metalline ore, he found first the iron-stone wrought in several ways, namely, upon standing passages and floats, fallings and proper pieces tinged, after the four Elements and colors of the Rain-bow. Then he considered exactly its flowers, according to the condition of each stone-work, how and out of what the iron stone may most conveniently be melted, and what manner of instruments may be used thereunto, where it may best and most firmly be wrought; for its ore affords a threefold society and wildness, which are useful , as namely *Glass-heads*, which are like a sharp blood-stone, breaking in a manner of a skull, are scaly, and brown *spissia*, somewhere of are white thorns, like the wood upon which *Abraham* purposed to offer his son *Isaac*. Secondly, the Brown-stone, out of which is made glass and iron color. Thirdly, granulate iron-filings in the float-work, which is so hard, that it can scarcely be forced to be gotten off, or be brought to right, and when the iron-stone is come to its perfectness, then it breaks off by piece-meal through the stone and rock, that there are found whole Mines of iron-stone, such is the iron ore in *Styria*. The best iron-stone is black, or red-brown, sometimes it inclines to a yellowness, some is of a cherry-brown in the floats and stocks, some are black and small *spissie*, some yellowish, which glitters among the rest, like a copper stone of a brown black marble, and of a fair glass, some looks like separated float-work, throughout the whole Mine, some is cloddy and hoary in clayish fields, which only is called the Driving, is as the sand-stone, most hurtful unto gold, because it affords most of the stacks, and very little of iron. Some sticks in the gray clay, which affords most malleable iron, but is of a brownish color. There breaks also good iron stone in tartareous and limy mines, and the most running is on the standing passages, in crifty sandy *Dalk-Stones*. The gross cliff stones break some in their flats. It usually breaks also in the fore and after Mine-works, where some of it lies off-washed among the Roasts, like a brown arch; and on the day there is no ore so common as the iron-stone, because it assumes and takes in other ores, and sets it thorough, thus often it changes its color and nature, after it there ensues *Glass-heads, Emasites*, brown stone, *Osemund, Bolus*, together with the red oaker and iron shell , all those assume the Nature of iron, and the iron-stone receives the highest metals, Gold, Silver, Copper, Tin, Lead, whereby it grows untoward, but gold and silver are not hurtful unto it, they make it malleable; that which is mixed with copper, or with other poor metal easily falls asunder, is brittle, of the same condition is iron-flint, producing out of many passages a huge flint, partly porous, like unto a black flat, which besides the iron-stone yields another grosser or more subtle iron. By this exchanging *Tubal Cain*, the great and first Mine-master did perceive, that the stones have their activity, he looked about, and finding that the Lime-stones, which contain iron ore, are of such mixtures, which may be burned to lime or calx, to raise walls with them; and know other sorts of *Tapff-stones*, as also calx stones are fit for to be burnt, and found them to be helpful for his melting. Thus, the iron-stone is associable unto other stones, be they metalline, or mineral. At *Musbach* there is copper shot iron, which has a lead joining thereunto; Founders must be expert to deal with such ores in their melting, and Magistrates do wisely that train up their subjects in such ways, for the good of the public. Thus is the iron the first and last Mine-work, a chief metal which many creatures cannot want it, being of a most necessary use, whereby things within and above the earth can be forced, no man is able to remember all the uses it may be employed unto, for everyday things fall out, to which there is need of the use of iron; iron easily receives a malleableness in a transmuting way, of which some of the ancient Philosophers have spoken; our iron is drawn from the Magnet, performs many useful works in the affinity with copper, which it is near kin unto, as also unto the gold and lead, for thereby are made the most glorious *Alkali*, which appear helpful in many things unto other creatures, as Poets write of, and attribute many strange qualities by way of parable unto iron; and if in writing all the virtues thereof should be comprehended, it would rise to a great Volume; its stones have in many Countries decreased, all other metalline stones are upon their decay, only gold, silver, copper, and lead keep their multiplying condition all the world over.

Chapter 7: Of Lead-ore
Its Mine, Condition, and Striking Passages

The lead ore is wrought under that heavenly impression of the black and cold *Saturn*, by an undigested waterish Sulphur, impure metal and salt. First, generally there is wrought a brittle glittering lead-color in that ore which is called *Glasse*, breaking in many rocks, containing gold and silver, yield gross and lasting Mine-works. Some lead stones are very broad, because glassy ores are mixed with it, with flints or marcasites, partly they are glassy, red goldish, white goldish, silvery, copper glassy, and of copper. Some lead ore turns to a blue color, mixed with a white transparency, like unto a shot *Bolus*; some is like unto the stone-salt and allome; some are of a dark green, like unto green floats, which lye gritty in a yellow or glue-colored clay, some are of a brown black, some are yellow red, like *Minium*, some are pure and compact, some are insprinkled and moving, some are mixed with iron, some with silver and lead, some are mixed with marble and flowers; some break also upon standing and level moving passages, and some are wrought in pieces here and there in slate-mines, where black lead lies along through the whole Mine; some are glassy in limestones, and some are very rich of silver in huge marble passages. There is a twofold Marble; the silver passages have a subtle light, and glassy brittle Marble, which looks like the glass upon gold Mine-works, is of a curious white glittering quality.

Lead-ore is wrought several ways, and the color of it changes after the manner of the ores, especially is the sorts of glass ores. For if *Saturn* lies below, or is in subjection unto others, then the glass has no power to bring *Saturn* unto *Saturn* an imperfect mineral, which either is too hard, or else untoward, and the *Nodus* of *Venus* is a *mispukel*, or a mixture of lead and silver, which is knitted very hard, but if soft, then it is water lead-glass, of the which are found in gold juices and tin-ores a kind of iron glass or iron mole, but is heavier and more brittle than iron-glass, by reason of its terrestrity or earthliness, which keeps in the metal, and is neither too soft nor too hard, and is glassy, white goldish, red goldish, and falls into the best metalline ores.

True lead-glass and ores afford half or the third part of lead, mixed with some other metal, and if one of the other metals be found in the glass, which keeps the predominance, then lead-passages are simply good, and lead is united with gold, and these are mixed stones, for the stones of mine-ores are more wonderful in their singular accidents.

Thus, is here the lead also in its fall, and bleak, after the heavenly impression which the highest has so endued, that it is subject to other metals, and is the supreme Finer in the essential Fruits of others. It easily mingles naturally with other metals, and the qualities of other ores, together with the leaves, bulk and roots into other stones of earth: And *Saturn* in his degree and power is the Highest, in a singular division of all his works, in which he shows himself in a clarified transparent soul, running into Antimony with its sweetness, which should merely embrace the gold; this is done so, not without cause; for in its ponderosity it yields the lightest remedy to all melancholy and heavy blood. As heavenly astrals are several, the clouds under them are of all sorts of colors; so the one lead is purer and more malleable than others, as that in *England* and at *Villach* it is seen in the lead-stones also. For lead-ores which are mixed with other stones, especially with such as contain silver, iron, copper, yield much of lightstones, and lead-work, which are picked out for separating, and the rather if they are rich of gold. Such worthy metals there are in *Hungary*, less pains are taken about them in their fining. Mineral flints with their unripe juices in the weak joinings of lead ore unite the *Saturnal* glass; if without any mixture affords to *Potters* a green glazure, if all be not melted into lead; but if you get a brittle mixed flint, there the glass is half upon iron, and such that are most pliable afford melting glass for fining for such sturdy wild ore, which will not melt. Artists may prepare such Saturnal glass mingling with it a small quantity of metalline flower, which will look, as fair as if it were a natural one. There may be extracted from lead an effectual medicine for mans' health.

If slate ores are found with another mixture, there are generated most fixed and firm copper, Vitriol and calamy also, as they are at *Goslar* in *Harlynia*. The best lead is in *England* and at *Villach*.

Man cannot well be without any of his members; metals, according to Gods ordinance are of the same quality, if man knew to make good use of them, for Nature has provided richly for him in that way; if men work these ignorantly, what utility can they have of them? Of the metalline soul is made a chain, which links together the junctures of gold and silver: these are endued with a special spirit, which is distilled into a water through a transparent head; Nature congeals under ground in the passages such water into ice, for a sign, that there is at hard a vein of lead, and silver, or of pure lead, and if there be a mixture of other metal about it, it is the better.

The best lead passages are such waters, blue, scaly, Talky, slate-stones, and fluid streaked marbles at length, or curled insprinkled ones, and not wrapped or wound about, with moving passages, almost unlike unto silver ores. Some lead ores are of a white, scaly, Talk-slate, full of white garnets, in which lead ore doth appear, which is rich of silver.

Chapter 8: Of Tin
Its Ore, Operation, Mist, Stocks, Floats, Fallings, and Striking Passages

Tin ore is wrought in a sand-stone, having its influence from Jupiter above, wrought of a dark brown, purple colored, grayish, black shining mercurial salt, and some sulphur mixed with it, interlined with an unkind gross sulphureous fume, all these incorporate together, making up the body of Tin: this unkind fume is the cause of the brittleness of tin, and makes all other metals that are melted with it unkind and brittle. This Tin or *Qwitter* grows or breaks in a threefold manner, viz., it slides, it is full of fumes, and it grows in pieces; it has a threefold wildness also, as *Shoel*, flint, and iron-mould which causes lead-work; their colors are black slate, brown, and yellow. These sand and *Qwitter* ores are environed, or enclosed in mighty broad standing passages, which appear to the day with *Qwitters*; some contain also rich paint work, some of these flints must be calcined, some are mixed with store of *Talk* and *Cat-silver*, which is a food unto *Qwitter*, and loves to stay there, some there are which grow in a *Glimmer* or *Cat-silver*, and is iron-mould, others also do strike in a fire stone or flint, so that fire must be applied thereunto, others are in a soft stone, and as it were swimming along. Some are richer than others. That which grows pure, and in black small stones, and is heaped together that natural work, that gives the greatest gain. And because *Jupiter* is the potent Lord of it, therefore it has a mighty throne and seat, that is, a mighty huge Mine-ore, out of which Tin is made by heaps, as is of that nature and property that it presses outward, and blossoms to a day, thrusting off Soap-work, whence come the wash work of Tin-Soap; for *Qwitter* does not grow in the sand of earth, besides in its body it is removed further from the seat of its throne to the foot-stool, making for itself a twofold dominion, in one it borders and reaches to slates and other stones that lie about it; insomuch that his dominion increases, in which is not a little, but much, on the blue stones, fallings, floats, passages, *Shools*, and cliffs, which incline one upon another, and do join, many times a mighty Tin-stone is wrought, which sinks down among its own cinder and slate, and at its sinking purges itself, and there come other fumes like clouds, which at all sides shoot into, and then breaking again as good as ever it did before; and it is of that good condition, that it despises no lodging, nor passes by any, but as poor and as despicable the stone is in that place, be it red, brown, fresh or stale, broad or small, it will press into, and mingle itself with it, and not be forced out of it, making itself great, little, gross, mild, tame, subtle and pliable, even as the rest will have it, and all this in a natural way; it loves to border upon silver and iron-stone, that Tin and Iron be united in a mighty fixed silver and copper ore, all which are found at their several marks. Tin ore is in this place better and malleable, if found afar off from flint-passages, and are less mingled with iron-mould, especially if copper stones, which in calcining can hardly be separated, proves Lead-work, without any fair glass.

Some of it is so mild and soft, that when they are cleared and calcined, still lose something, for flints and sulphureous matters, which are volatile, and cannot endure any great heat, corrode somewhat of the metalline Tin, which appears by the white thick fume at the calcining: they are calcined thus hard by reason of *Bake-iron*, else they might yield as much again; for they lose extremely in calcining. It is strange to some, why they shrink together to so small a quantity, being they get a greater quantity of lead with good *Qwitter*, at first brought out of the Mine.

Chapter 9: Of Mercurial Ore and its Passages

Mercurial ore is wrought in its proper mine-stones, by the quality of its salt earth, and its nimble volatile earth, in a moist, greasy, slimy, waterish *oletity*, which is mixed with a most subtle, red sulphureous digested earth, with a most weak slow binding, like an unripe pleasing fruit of all particular metals.

Mercury shows its virtue in many things admirably, and works effectually upon minerals and Metalline sulphur, and upon such which border upon Antimonial stones or ores; it loves to be in such places where the Tin-ores lie higher than silver passages. It requires many iterating effectual operations unto other ores, and is multiplied upon other strange stones, and is drawn through the juices of Minerals and Metals, which are in affinity one to another, and produce many strange miscreants, this is the reason why it is so pleasant unto metals, Goldsmiths amalgam and gild with it.

It is used also for metalline colors, and is prepared to an oil and water, for mans' health, and is sublimed for to corrode this worst of poisons, and is a true Robber, taking along whatever costs have been bestowed on him; but if he can be caught in his nature, then is he in subjection, and obedience unto quick and dead. He is very effectual, in *Medicine*, especially for outward forces, he is naught to naught, and good to good, and is not every bodies friend, though he is willing to do what you put him upon. His metalline stones are of the same nature with pure white slate earth, inclined to a water-blue, in fresh intermingled white marbles, in a glassy grayish and porous *Glimmer* or (cat-silver) which lie beneath betwixt the slates, in a float way, which are mingled in their metalline passages within-sled *Marcasites*, and with the subtlest small streaked white Talk, and are thorough grown with two sorts, standing and float-striking passages, in which is wrought a curious red shining quick-silver ore, not unlike unto red Mine-sulphur, and sometimes flows purely out of the cliffs and caves of the passages, stands in a sink or puddle together like water, which its natural quick substance sufficiently evidences.

Chapter 10: Of Wismuth, Antimony, Sulphur, Salt, Salt-peter and Talk

Wismuth is wrought in its own Mine-stone, not quite freed from a protruding silver, or Tin-stone, of an imperfect pure quick-silver with Tin-salt, and fluid silver-sulphur of a brittle immiscible earth, partly of a crude fluid sulphur, partly of a mixed much exsiccated sulphur, according as it has gotten a *matrix*, after it was conceived: then it turns a bastard of a brittle nature, easily unites with Mercury, and is wrought naturally in a twofold form, the one is fluid and metalline, is melted with dry wood, being mixed with clay, yields much of white *Arsenic*. The other is small streaked, or *spissie*, remaining an unripe substance, yields a fixed sulphur instead of Arsenic: both these are silver *Wismuth*.

Antimony comes from perfect Mercury, wrought of little salt and a waterish fluid sulphur, though it shines black naturally, and its out-side is of an antimonial form, yet it graduates and purifies the noble nature of Gold, and does much good unto men being artificially prepared in its several ways; notwithstanding its color, it keeps its high and mighty praise and virtue: For meeting with a Master, which can clarify it, and gets its natural Gold out of it, and extracts a blood red Oil from it, that serves against many Chronical diseases; it must be reduced to a transparent glass: this black evaporated unripe metal represents to us Gods Majestic glory, who is not a regarder of persons, bestowing upon poor despised men rarities of virtues and knowledge!

The red mine-sulphur, which is found in *Tyrol*, *Tonawitz*, and *Engadin*, and grows in a black blue flat-stone, and has singular innumerable good virtues, wherein lies hid a mighty purifying quality, lies on with its color unto the red goldish silver ore, or Cinober ore, and looks almost like unto it, whose redness shines forth most pleasantly.

Salt has its special virtues to penetrate and to preserve from putrefaction, contains a noble spirit: and it were very necessary, that men would not be so careless and neglect in their seasoning with salt, suffering matters to stink and corrupt,

112

considering too slightly, and taking so small notice of the noble gift put into good mineral works, better lying on their hoary old walls.

Talk is an ingrown sulphur, shines incombustibly like gold and silver, closes and bows, is transparent like glass, is called *Sulphur*, *Lutum*, keeps in the fire incombustible, like *Alumen Plumosum*, lies in rocks and Stone-works, serves for graduating of metals. Every metal, mineral and salt in particular is good to be used, each is distinguished in its particular name: even as those that make glasses, put their several names upon them, and put their several forms upon them, making them into drinking glasses, flagons, bodies, bolt-heads, helmets, receivers, pelicans, jar-glasses, wine-glasses, funnels, all these he framed after his own fancy, either into small, great, long, or round forms, even as he pleases.

Chapter 11: A Comparison between Gods Word and the Minerals

Like as the heavenly glorious God in a Spiritual way, in his most dearest Son our lord *Jesus Christ*, at his redeeming of mankind for the good of man appeared a Sun of Righteousness, which glory the Prophet *Esaias* has prophesied of in the Lords Spirit many years ago: How two *Cherubims* and *Seraphims* having six wings, moved and sung before the Lord: Holy, holy, holy is the Lord Zebaoth, of whose glory all the world is full, which Prophet has seen the most omnipotent Lord of Lords, knowing him a God in a Triple essence, and that out of that noble *Chaos* of *Jesus Christ* should flow the fountain of life, of mercy, and righteousness, which the Lord God made apparent on the Tree of the Holy Cross, where out of the side of his dearest Son did run Blood and water, to which the Lord in the Revelation of St. *John* adds, fire, smoke, and fume; this union according to the Divine Word is grown at the beginning in all creatures, and whatever God the Holy Trinity has ever created consists in a Trinity, even as the Deity is in an eternal Trinity: As the Deity is indivisible in the Humanity, α & Ω, in the water and blood for an eternal remembrance, that is, the first and the last letter: as in the Heavenly, even so in the earthly, the perfect Alphabet must not be cut asunder, all must stand from the beginning to the end; and Christ Jesus purges his dear friends still unto eternal life through water and blood, saying to their hearts, all thy sins are forgiven thee, thy faith doth save thee. No man is saved, unless he be first born again, that is, through water and blood, which thoroughly purges not only men, and the sons of men, but also the whole *Limbus* upon earth; for it is not the metalline blood and water, neither is it Mercury and Sulphur that does it, neither in the body underground is any goldish silver wrought to any blood red ore, the blood out of Christs side shed for the good of man, is that great evidence for thus all Mineral stones, that are in the plain element of earth, and the spirit of all ores, and marbles, and stones come from the divine essence, as also the heavenly spirits for the throne of God, with the heavenly Angels and Spirits are furnished for the praise of God: thus the earth also is created in her stones, ores, veins, passages, for the honor of God, and the welfare of man, which imitates Gods wisdom, filled with infinite and unceasing forth-bringing of fruits.

Whence should be the decay of metals? Surely even as the eyes of the holy Apostles and Disciples were held, that they could not know the Lord in his clarified spiritual body and essence; no more can men see the things in metals.

Why does Saint *John* in his Revelation speak of smoke and fume? Surely he did not mean the fire, smoke, and fume of Bakers ovens, or Kitchen chimneys, but there was revealed unto him the heavenly fire, the mist, vapor, and fume, which is exhaled from the moisture of earth, and elevated to the clouds; so in the subterranean works the fume and spoil, or outside of the ore are sublimed, and the fire of the frost which rouses the effectual powers, vapors, and spirits makes then come to a perfect unity in metalline bodies. Now if there were not a fire and vaporous fume in the earth, how could they produce their fruits, which are the minerals and metals underground?

As the fiery element is covered with the airy, and the heaven with clouds, and the earth is filled by them, and together with the fire was enclosed as one element with the other two. In like manner, at the first Creation, the subterranean passages and veins were laden with ores, as trees were with fruit, which the Lord God in Paradise had implanted into them. This effectual fire, vapor and fume is likened unto Mercury, Sulphur, Salt, and Sea-water, wherein earth lies enclosed and hidden even as the supreme throne of God is encompassed by other thrones and heavenly habitations.

As the four Evangelists are witnesses of the New Testament and Covenant; so they are a type and sure testimony of the four elements, that the earth is created after the holy Heaven; thus, we are taught in the Lord's Prayer, as it is in heaven, so in the earth, in which, and beneath, and under God is everywhere. This is in action still, King *David* could confess, that he could not hide himself from the Lord anywhere.

Seeing the holy and blessed God has laid the creatures in the earth with the four elemental qualities, therefore let rational Miners open the eyes, and learn judicially to know the passages and cliffs of ores, metals and minerals, then they will get a lasting name with great praise, and will be like the noble gold, which in a glory and beauty appears, when it comes from the *Quart*[14], and can be then reduced into an oil, which preserves man in a lasting health, beyond any balsom, and is become a vegetable, which is potable.

It is feasible, that of gold may be prepared a singular medicine for the good of man-kind, because man is created of God from *Limus terrae*, and the whole earth is a *Limus*, such another medicine all the Doctors are not able to produce, which is of a curious sweet fragrancy, standing distinct in two lights, and must needs be in *rerum*

---

[14] Exactly as in the 1671 edition. -pnw

natura, because it was brought on God his Altar, for an offering by man's art prepared, suffered it to be extinguished. None know what it is, neither the literate Doctors know the preparation of it, who when their Confections, Syrups, Herbs and Potions will do no good, and are in despair, then they might willingly run to Metals, which formerly they made conscience to make use of them in their Ointments and Plasters; of this I make mention in a reverend remembrance for true rational Miners. Out of gold and silver are joined not only gold and silver monies and other plates for man's use, but they serve for man's use in many other things; and offer the first metals virtue, there come others also more and more very effectual, even to the last of metals.

Such virtues there are in minerals also, as in vitriol, antimony, allome, salt, and the like. All these are a nourishment unto metals, even as Manna was to the *Israelites* in the desert. As they are easy withdrawn and taken from metals, so it happened to then also, Heathens and Christians received that Manna, together with Mines and Kingdoms, they are set and shot at the heap of rubbish, where they still worship the Calf; of this I have spoken more in that book where I treated of *Fossilia*.

Chapter 12: How Precious Stones and Jewels are Wrought And How God has Bestowed Blessings Upon Those That Work the Mines

Jewels are wrought out of the substance of the most perfect, transparent, and noblest earthliness, with a mixture of the noblest Mercury, Sulphur and Salt, without any fume, or moist matter: are a dry coagulation, and commonly are engendered in a round form in their dwellings, lodgings, stocks, and passages, fixedly bound together, some are of a transparent luster, others are more dark; and they have their several colors.

Not many ores are found, in which these noble generated bodies are brought to any perfectness, neither are their strikings along in a way of passage, here and there they have their own Centers, into which are joined tender and miraculous accrescencies, where they are *guttatim* lapidated, falling into hardest, purest stones concavities, growing in several cuticles, as we see the animal stones do grow. The more precious the Jewels are, the fewer there are of them; and the grosser their mixture is, the more store there is found of them, which is apparent in *Garnets*; who has hitherto searched into the quick spirits of such noble creatures, the Lord has created for man's benefit?

Pygmies or *Homunculi*, which in former times lived in hollow ores of mines, these could not want skill in such ways, having traversed and travelled up and down all these slippery corners and ways. The places and situation of such Jewels lying somewhat nearer unto Heaven, in the Eastern Countries, bordering on Paradise, so there must needs be abounding in Gold and Jewels, and such precious Vegetables, which our thoughts hardly may reach unto. God requires no more of man, whom he entrusts with these things, but to be faithful and just, and is an argument for us to think that for the same cause pious Kings and Princes, and the old wise Patriarchs were gifted from above to bear a love to search into Mine-works, and did it with an uprightness and judgement. Let honest godly Christian Miners choose the better part, and learn to know the pearl, the Spirit of the Lord proceeding out of Gods own mouth, and let them consider well their eternal fixation, to return their love again to him that has loved them first, bringing all things to their subjection, he imparts all unto them abundantly in grace and mercy, and by the innocence anal merit of his only son, bestows on them temporal and eternal blessings, and puts more glorious ornaments on them, and better than ever gold, silver, jewels, and pearls were adorned withal.

Chapter 13: Of the Essence of Gold
Abundantly found not only in the Metal, but Mineral also, Whose Energy is Served Most Rarely, and a Short Closing of my First and Second Part of Minerals and Metals is Annexed

This Chapter is a breviary of all mineral colors and forms, how they after an heavenly operation are daily clad in the metalline prime *matrix*, and set forth in their several words, whereas there shines forth unto us the eternal light of the lustrous Sun, the deity of the day of joys, and of the eternal most fixed and fairest *Sol*, as also of a most yellow, pure, red, and fixed citron color of heavens eternal lightning and the most glorious paradise of all the Stars, a natural created light for all creatures, besides the beautiful *Aurora* of Mineral Earths, and of their subtlest, compact, and best binding enclosed, speaking to all other white untinged metals; I, *Sol*, of an essential being, am Lord of Lords, in power, might, and perfection, I overcome all , and I overcome and bring them into subjection, and none of them can master me, but I do conquer them all, they are subject to me, and to my Being, for my Kingdom is established with infinite and invincible Power and Dignity; by me all metals, minerals, animals and vegetables are strengthened and rectified; for I give to everyone that knows me in my green, blue, and red Nature, all what I have, and what he desires, I cause to drop down after the four cardinal streams of *Pison*, *Gihou*, the noblest substance of Mercury, in the form of a most pure transparent crystalline water, and the most noble substance of Sulphur, of *Hidekel* and *Phrath* the clearest fairest Astral salt from a Vitriol salt, which through all mines flew upward very fruitfully, and penetrates all the mineral stones. I alone graduate and exalt the silver, unto *Luna*, I give light and luster in all righteousness, of my virtue do speak all *Magi*, Naturalists, and Scribes all the world over, from the East to the West, I am the Lord over the heavenly clarified garments and colors, I adorn the firmament, the weather, clothe the Rainbow after Gods will, I exalt all jewels, all such growths and creatures and what I cannot inwardly walk through and reach into my course, I leave it to be perfected with my friend and lover the *Luna*, she receives the best part of me, and of the subtlest an abundance, the *Indies, Hungary, Carinthia* testifies the same, for all what is to live, and is to receive a life, rejoices in me, and next to God, in none else, for to him honor and glory belongs solely after him, I find no higher Lord and Commander. But for my part I do not rest, neither do I desire any rest, do my office readily into which my Creator has placed me, I let my pliableness be found gloriously, like a wax in stones, which have by reason of hardness fire enough, if need.

I am hidden from unwise men, and am ready to be discerned by men of understanding. I am predominant abundantly in a well-known mineral, as also in *Mars* and *Venus* which are of low degrees: in them I lie hid also, all these have a double spirit, well known unto *Lune*, pleasant to her, and next unto her. Hence God suffered *Moses* to erect a brazen Serpent, in the desert after my color, in harkening unto the people, at the Mount *Sinai*. My best and fairest color appears in transparent juices, as vitriol, which after my condition in due time penetrates ores, whereby they grow rich in lust, and are trained up in a pleasant form corroded into a greenness, like sealing wax, green like Goose-dung, blue like Saphir, and so forth, sometimes of the color of a water-flint; by red and white color is the best, which are heartily wished for. I love to be kindled in vitriol, and further it after descension in its green food, into a deep red spirit, after whose laxative purging comes that expected *aqua Saturni*, the true *acid-Well*: from whence I myself and all other metals, animals and vegetables have my offspring and life. For Metals and Minerals rise only from thence, have their beginning and original from it, for it is that quickening water, which ordinary Miners do not know of, is known only to Philosophers. It works Minerals and Metals in several ways, in form of taps which did skept, pure, white, compact, sound like purified Sugar, in a blue slate-work. An extraordinary pleasant mineral for all colors. Salt Ores are at a further distance, which by my attractive changing, are found in floats, blocks and passages, which in many places, bring the water unto the day-light, so that it often is found a pure and dry Salt above ground, of glassy light flames, or in a great frost like unto flocks of snow, there shoots a brittle, glassy light stone, wrought in great pieces: in the same order are all other Jewels according to mine enlightened heavenly stone, distributed among their operations, worths, and virtues, and clarified in a most fixed transparency, and endued with an everlasting spirit, distinct in several colors, as Diamond, Smaragds, Carbuncles, Saphirs, Rubies, Crystals, Chalceodonian, Jaspis, Berill, Chrysolith, Onyx, Carmel, Turkois, Lazur-stone, Margarits, Corrals, *Terra Lemnia*, Terpentine-stones, and Garnets, of deeper low colors, each in its heavenly colors order is transparent, and naturally is created, and preserved in its own *officina*: Hence it may be argued, that all these together with good fruits serve for man's good, both for his body and spirit; for nothing is hid from my transparent power, my splendor and luster over-shadows all these, and are held to their growing unto maturity: let no creatures marvel at these several distinctions from whence they all should come, for all have their principle from me, and from my spirit, which is hidden in me, which none can dive into, save the sole Creator of all things, from whom it proceeded as out of his Divine mouth. Thus, I close up my speech, and myself startle at so great a mystery, and attest in truth for a farewell, that I am not only the Gold and present *Sol*, but give also strength and power to all the inferior terrine spirits; for *Aristeus* and *Onizon* is in subjection unto me, for I am α & Ω, God be praised forever.

Thus, I conclude the second part of my Mineral book, wherein I have showed faithfully as much as I know, and could in my industry apprehend: let others do their endeavors also, let them produce their knowledge also, that the light of the noble nature may still be supplied in her plenitude, and may not go out, whereby cause would be given to the enemy and envious men, to be outrageous against such truths. Let God still and incessantly be importuned with prayers and thanksgiving. For these ends I have written these my two Treatises, and annexed the manuals at the beginning (which otherwise needed not to be done) that by earnest prayer and thanksgiving, and continued earnest worshipping of God, every one might carefully exercise himself therein, and be convinced in his reason, how gloriously almighty God has created, ordained, and held forth nature, to perform her operations underground, and to produce unto the day light formally their Nativities and fruits, that we may reap thereby not only our sustenance, but may acknowledge Gods infinite mercy and goodness, for which none can return sufficient thanks. However, let everyone do his duty, and as much as he is able to perform with his heart and tongue, pray to God in sincerity for his grace, blessing and wisdom, to conceive by his spirit of truth and righteousness of his great and wonderful Creature, that the honor of God may be exalted above the Heaven, and be proclaimed with infinite praise throughout all the World.

END OF THE SECOND PART

THE THIRD PART

OF

BASILIUS VALENTINUS

HIS LAST TESTAMENT

TREATING OF THE UNIVERSAL WORK IN THE WHOLE WORLD, WITH
A
PERFECT DECLARATION OF THE XII KEYS: WHEREIN IS
SIGNIFICANTLY EXPRESSED THE NAME OF THE GREAT MATTER.

THERE IS AN ELUCIDATION ALSO OF ALL HIS FORMER WRITINGS,
PUBLISHED FOR THE GOOD OF THE POSTERITY, AND SUCH, THAT
ARE LOVERS OF WISDOM.

LONDON

Printed by S.G. & B.G. for Edward Brewster,
at the Crane in saint Pauls Church-yard.

1670

The Third Part

A Declaration of the XII. Keys

Here follows the third part of my intended writings, wherein is truly showed the original and prime matter of our Philosophic stone, which is a perfect instruction to the practical part, which shows the direct way to the inexhaustible fountain of health, and of the abundance of riches to provide for man's necessaries: and this is a Declaration of my former writings, which is left for a love-token to all those that are lovers of Philosophy.

My friend you must note, that by this intended work touching the Minerals, I had reason to prefix the two parts of Minerals, and Metals, and their Ores, holding it a necessity to hold forth a light unto the ignorant, how that one spirit from above frames all such Ores, Metals, and Minerals, taking their original underground for to generate thereby. For earth is always ready and covetous to attract and to retain that spirit, which proceeds from Heavens powers, which it presents in process of time in a formality and perfection. The matter of it has been spoken of sufficiently in my former writings, which is the reason why I give only hints of them in this place.

Note, that all things proceed from a heavenly influence, elemental operation, earthly substance, from this mixture arise the four Elements, water, air, earth, which engender by the help of fire hid therein, in a warm digestion, producing a Soul, Spirit, and Body. These are the three prime principles, which in a coagulation come to a Mercury, Sulphur, and Salt, these three being in conjunction, according to the nature of the seed produce a perfect body; be it in the Kingdom, either of Minerals, Animals, or Vegetables. All things in the world, that are visible and palpable are divided into these three Kingdoms: The Animal which contains such that have a lively breath, composed of flesh and blood, as men, beasts, worms, fishes, fouls. The Vegetables, which contains trees, herbs, seeds, roots, fruits, and all such things that are of a growing quality, the Minerals contains all manner of Ores, Metals, Minerals,

Marcasites, Calxes, Zinks, Lobols, all sorts of flints, pebbles, wismuths, stones, precious ones and others.

Animals have their special seed, a spermatic substance, which after copulation generate flesh and blood, which seed is their *prima materia*, from a heavenly influence, created by God of the four Elements, wrought by nature, which formerly were quoted in my writings.

Vegetables also have their proper seed which God bestowed on then, according to their several qualities and form by a heavenly and sidereal influence, and receive their elemental fruitful growing from the earth, with an order, thereby to generate and augment.

Minerals and *Metals* also have their original seed from God, by the heavens influence in a liquid aerial substance, by a Mineral spirit, sulphureous Soul and earthly Salt in one body joined: of these I have spoken in my former writings.

Note further, if any of these Metalline and Mineral kinds shall be brought to a further propagation and augmentation, it must be reduced to its first seed and *prima materia*.

If you will transmute Metals, augment them, bring them into a tincture, or Philosophic stone, you must first understand, how you may destroy by a Spagyrick Art, the Metalline and Mineral form and separate it into Mercury, Sulphur, and Salt, these three must be purely separated, and brought to their first principles.

This separation is done in, and by a Mercurial spirit, sulphureous Soul and white Salt. These three in a due ordering of a true manual must be joined again, that they may be brought to the highest and most perfect purity.

In which conjunction must exactly be observed the quantity: after this conjunction the whole substance is merely a liquid substance and philosophic water, in which all the Elements, first the heavenly, then the elemental, and lastly the earthly qualities are shut up and lie hid therein.

For the Mercurial spirit is cold and moist, the sulphureous Soul is warm and dry, and this liquor is the true *prima materia*, and first seed of Metals and Minerals, which by Vulcans Art is brought to a *plusquam perfection*, into a transcendent fixed Medicine, out of which is generated the true Philosophic stone, and must be produced in that way.

Therefore, observe and take notice, that all Metals and Minerals have only *one root*, from which generally their descent is, he that knows that rightly, needs not to destroy metals, to extract the Spirit from one, the Sulphur from a second, and the Salt from a third. For there is a nearer place yet, in which these three, Spirit, Soul, and Body lie hid in one thing; well known, and may with great praise be gotten, it shalt be nominated afterwards in several terms.

He that learns to know exactly this golden seed, or this MAGNET, and searches thoroughly into its properties, he has then the true root of life, and may attain unto that, his heart longs for. In my former writings, as also in the **XII. Keys**, from the first to the last, I ordered thus my style, in writing, wherein I held forth unto posterity the *practick*, how the great stone of Philosophers, or the best purified gold may be made out of Sulphur and Salt, with the help of the Spirit of Mercury, which must be drawn from a crude unmelted *Minera*, according to the tenor of my **Fifth Key** set down in a parabolical manner.

Why I did the work of the Philosophic stone upon Gold Metal, this is the reason that the simple Laborators, to whom is unknown the other body, or *Subjectum*, which contains all the three principles: though it be a thing well known, yet is it a stranger to their brains, may learn hereby more wit and knowledge. Many of the ancient Philosophers, which lived long before me, have in the same manner with me obtained the true universal stone of all mysteries and health, as their books, which are extant, give evidence thereof. The first time I took great pains and was at great expenses, and consumed much time about the purified Gold, alleged in the **First Key**, this heavenly stone I prepared in the Cloister I lived in, and happily obtained it. The highest in heaven bestowed his further grace and blessing upon me, that I took unto further consideration the tinging animated spirits placed and planted into their several bodies. Let no man be ashamed to learn, to add more to his learning, and to dive further into that, which was hid from him, notwithstanding his knowing ways. Nature reserves many things in her secrecy, which men's dull understanding and shortness of life cannot reach into.

Whereas God in his great goodness has bestowed this great gift upon me, for an improvement of that talent, I have imparted the same to my fellow Christians in the said XII. Keys.

Those that are endued with deeper wits and knowledge, and in their hearty and careful endeavors strive further to dive in the Art, will meet in the same place with a more easy and more known matter, which was almost named and set down, of an effectual quality, out of which in like manner, as the ancients before me, in their exact speculation and practick have in the end better known the only scope and drift, which has been practiced several times by me also, in a shorter time, and less pains taking, both they and I have obtained health and riches: in this known and despicable matter and Mineral substance is found a Sulphur and tincture more effectual, and more worthy, than the best Gold can afford, which is fluid and open, and its Mercurial spirit also, and its mystical Salt is free and open, whose virtues may with less pains in a visible manner be drawn from it.

He that has considered exactly my XII. Keys, frequently perusing the same, most needs conceive, and that therein is held forth the whole preparation of our stone, from the beginning to the end, without any defect, yet so that it only should be prepared of Gold fitted for it. But we, according to Gods ordinance in nature, have pointed at a Gold, which is much better, and requires to be taken into a deep consideration, which being unknown and strange to Novices, for some reasons I forbear to give them any direction thereunto, with a resolution to write and to point at such matters, as themselves are inclined to seek for their seed in them.

At the beginning of my XII. Keys, according to the manner of Philosophers, in a parabolical way I made relation of the property and work of our stone and balsam, how it was made by Artists, which as by an inheritance is come to me also, wherein I spoke as much as was meet of the government of the fire, chargeable appearance, and of the chiefest planetary colors, and the final end thereof. After the accomplishment of these peruse well the XII. Keys, for each contains a *particular* work.

The **First Key** informs you, that if you seek for the seed in a Metalline Body, as in the Gold, then before all things it must most exactly be purged from all its impure leprosie, and that nothing must be mixed with our Fountain, but such, which is of a pure spermatic quality. This purifying is performed with Antimony, which stands in a near relation, and affinity unto Gold, which is the reason, why antimonial sulphur purges the Soul of Gold, graduating the same to a very high degree. On the other side, the Gold can meliorate in a short time the Soul of Antimony, and can bring it to a firm fixation, exalting Antimony and Gold to a dignity and virtue, and can be brought not only into a white metal of *Lune*, but also to a transcendent medicine for man's health, of which you shall have a further direction hereafter, which I shall treat in particular of Antimony. Although Antimony has promised unto *Saturn* a sociable brotherhood, because Antimony's qualification does rest in some sort on the quality of *Saturn* in an equal concordance, yet after the fixation of the exalted Sulphur of Antimony, his next friend *Saturn* cannot get any prey from him, because the King received him into his golden Palace, and make him partaker of his triumphing Kingdom. This is the reason, why he can endure now heat and frost, and overcomes it, and stands with the King a Conqueror in great and transcendent glory.

The purifying of the Gold is performed thus: Laminate the gold thinly, after a due manner, cast it thrice through Antimony, afterward the *Regulus* which is set at the through casting, must be melted before the blast in a strong fire, and driven off with *Saturn*, then you will find the purest, fairest, most lustrous Gold, pleasant to behold, as much as the luster of *Sol* is. This Gold is now fitted to surrender its innermost, being first brought from its fixedness into a destructive form, and pass through the Salt-sea of its corruptibleness, is drowned therein, escaped again, and appears visibly.

The Second Key

My friend note, and take that into a serious consideration, because the chiefest point lies herein; cause a Balneum to be made, let nothing come into it, which should not be there, that the noble seed of the Gold fall not into a destructive and irrecoverable opposition after its destruction, and take an exact and careful view of such things, which my Second Key informs you of, namely what matter you ought to take to the *Kings Balneum*, whereby the *King* is destroyed, and its external form broken, and its undefiled Soul may come forth, to this purpose will serve the *Dragon* and the *Eagle*, which is *Niter* and *Sal Armoniack*, both which after their union are made into a *Aquafort*, as you shall be further informed of in my Manuals, where I shall treat in particular of Gold, of other Metals and Minerals, into which *Balneum* the King is thrown, being first, as in the quoted place you shall hear, brought into an *Amalgama* of Mercury and of Sulphur, which presently seizes on him, corrodes all his member, and is dissolved, and is presently mortified of this Salt-water, into a most splendent transparent Oil. You must note, that this dissolution is not sufficient, and the King is not minded as yet to let go his Soul out of his fixed body, which you can see when you separate the water from the dissolved body of the King, where you shall find fixed powder of Gold, out of which you will hardly get his Soul that is therein. Therefore, follow my counsel and bear the yoke, which I bore before you, and learn to know exactly in pains taking, further thus, as I shall inform you. Having dissolved your Gold wholly in the said water, and brought it into a pleasant yellow Oil, then let it stand well luted for a day and a night in a very gentle *Balneum Mariæ*, the *feces* which are settled, must be separated from it, then take this pure dissolution, put it into a well coated body, or Retort, apply a Helmet to it, with a Receiver, in the best manner luted to it, set it into a Sand Capel, drive the Gold with the water over the Helmet, iterate this a third time, abstract the water in *Balneum Mariæ*, you will find a fair Gold-powder, keep this in a glass for an hour in fire, let the remaining humidity be drawn from it.

The Third Key

Then take of good spirit of Salt-niter one part, and of dephlegmed Spirit of ordinary Salt, three parts, pour these spirits together warmed a little, into a body on the fore-written Gold powder, lute a Helmet and Receiver to it, drive the Gold over as formerly in sand several times with an iterated distillation, the oftener the better, let the Gold come to be volatile more and more, and at last let all come over. By this repeated driving over, its fixed body is divided, all its members are torn asunder and opened, and leaves willingly its Soul to a Special judge, of which my third Key will give sufficient information.

Note further, that after this work those salt spirits must be abstracted from the Gold, which was driven over, very gently in *Balneum Mariæ*, let nothing of the tincture of the Gold come over, that the body suffer not any diminution: then take that Gold, or rather these Chrystals of Gold, from which you have separated the water, put it in a Reverberating pan, set it under a Muffle, let its first fire be gentle for an hour, let all its corrosiveness be taken away, then your powder will be of a fair scarlet color, as subtle as ever was seen, put it in a clear viol, pour on it fresh spirit of ordinary Salt, first brought to a sweetness, let it stand in a gentle digestion, let that spirit be deeply tinged and transparent, red like a Ruby, cant it off, pour on fresh, extract again, iterate the work of canting off and pouring on, till no more tincture of it appears, put all these extractions together, separate them in *Balneo* gently from the Sulphur of *Sol*, then that powder is subtle and tender; of great worth; this matter is such, which in a short process transmutes *Lune* in its tincture to the highest perfection, according to the direction of my XII. Keys.

He that has some knowledge herein, may make this query: whither this extracted dry Soul and Sulphur of the King be just that Soul, of which Philosophers have this saying: the Philosophic work for the preparation of the most precious stone requires three things, viz. a wet volatile Mercury, or a Mercurial spirit, a wet volatile sulphureous Soul, and a dry Astral Salt, which after its dissolution together with the two first must be apparent and known in a waterish form; which way comes that about, because in this process nothing is spoken of any Mercurial spirit and volatile Soul, but the Soul of the King appeared in a subtle form of powder? The answer hereunto I delay so long till the Querist learns better to understand the distinction in this book, and I will perform my promise, and set his anxious and entangled mind at liberty, which is so much troubled about this doubt, and will deal with him as a good father may deal with his son; in and with this scope, wherein our Mastery lies, have been fooled most of the Wits, leading then captive in their erroneous ways, being led about in a desert of misled ways, because in their supposed deep wits, they had not conceived so much of the manner, how all things of the world are generated, and that every Spirit must have a Soul and every Soul a fit Spirit, and that both Spirit and Soul are spirits md spiritual, which must have a body, in which they may have a dwelling.

Gold and Silver, but chiefly Gold is brought to the highest fixedness, by such degrees as nature did afford, insomuch its nature is found very hot and fiery, freed from all phlegmatic humidity, of which *Lune* is not so wholly freed, though she has obtained a Sulphur-fixed degree, and stays for the King, to warm her cold body with his hot seed, which concerns the particulars, and belongs unto them, which in that place shall be plainly demonstrated. In Gold, there is no waterish humidity at all, unless it were reduced again into Vitriol, which would be but a useless and unprofitable work, and would require huge expenses, in case the Philosophers stone should be of Vitriol of Gold, of which there must be had great store; indeed, in that Vitriol there would be found a convenient spirit; which nature would desire, of a white quality, as also a Soul and Salt of a glorious essence.

But what Countries, Goods, Lands, have been dilapidated this way, I wave to discourse of only, this warning I give to my Disciples, nature having left a nearer way to keep and to imitate that, that they also might take heed to fall into such extreme and inextricable poverties.

The Solar Mercury *Sol*, being never brought so far unto destruction, neither did the ancient Philosophers ever make use of that way, as being a thing clean contrary unto nature, containing indeed a humidity, but it is a mere Elemental waterish humidity after its dissolution, and good for nothing, water and other principles do not stand in the Elements, but the Elements rest in the principles and seeds of Metals, of the which I have spoken formerly. Therefore let none be so over witty, as to make our stone only of dry and fully digested Gold; because its phlegmatic humidity is entered into a dry fixedness and fixed coagulation, which is not found so in other Metals, though they also are subject to a hard coagulation and passed through the fire, yet are not wholly digested, nor brought to a full maturity from the natural original root; which ought to be taken notice of; a:rd be not offended at my former writings if they seem to run contrary against this.

Though I have showed, that the Spirit, Soul, and Body come all from one Metalline essence, and must be prepared thus, among which I hold the Gold to be the best, however I dealt herein as it seemed fit for a Philosopher, the like the ancient Philosophers have done before me, but I hope you took notice of my protestation, that I gave special cause thereby to your speculation, to take the better notice of nature and her principle, and to consider the original, because it was not meet for me to inform all men, how the doors are bolted within, and especially at that time, when I never intended to write thus plainly of these things, which are hid even from the best of men, but when my heavenly Prince commanded, at the changing of my mind, not to bury the imparted talent, but to do the like to those, whom God think worthy, to leave it to them. One rule more must I put to thy remembrance, of such points, which formerly I have set down, of which I spoke now, that you may the less blame me, as if I did refuse these things now, which I formerly wrote of.

Peruse all such which since the beginning of the world have written of Metals, you will find, that they were all of one mind, and that I make use of their sayings:

that the first and the last Metal is a Metal, because the first Metal has already obtained, and gotten the forth-going seed of Metals in a Metalline quality, which does nothing else, but that it goes on incessantly in the Metalline generation, as I spoke of in the first and second part of Minerals and Metals, and in this part also I have spoken of it in several places.

Many have called Gold Lead, and Lead they called Gold, because it was found not only of the same ponderousness, but because three deep glittering stones have solely gotten from this Planet their transcendent perfection, and many other causes besides, which to relate here, would fall too long and needless. And this is it which asks wisdom to distinguish in this and other things, and exactly to search into Gods mysteries, and natures laid before us. But man, through *Adams* fall being brought to a deep blindness, therefore men's understanding is so eclipsed, that they can hardly conceive of this, and of other mystical matters in nature.

The obduration being so great among the covetous, that for the most part they search and dive into such mysteries out of mere covetousness, pride, and ambition, made the ancient Philosophers upon command and inspiration of the highest aim at that, as to put a certain stay to their hands, and to write such mysteries in such a manner, that unworthy men should not understand it, and but worthy men only in their illumination might perceive it; and writing often one thing , have mingled other among, understanding still the one and the same. In several places, they showed, that the Philosophers stone is, and comes from an animal, others from a vegetable seed, and a third sort says, it comes from a Mineral seed: others write that stone is made of an animal, vegetable, and Mineral seed together. All this is only understood of the Mineral and Metalline seed, and consists not in any plurality of seeds. Thence this Art grew eclipsed, insomuch that scarce one among many thousands hardly attained unto the knowledge thereof: and for that reason is it held for an Art, because not every Dunce should bring it into his Beetle-head, and why should it? For if this Art were as common as brewing beer, and baking bread, then any one may judge what good could be looked for; would not all manner of vices be practiced publicly without any controlling?

Therefore, such men, which in their lofty mind are merely for Pomp and Pride, must be clipped in their wings, and these things ought not to be put in their mouths, things are clear enough for those, on whom God intends the bestowing of them.

I return now to the thing I intended, which is to teach a desirous Scholar, how to proceed further with the extracted Soul of Gold: Truly it is much to discover such mysteries, I warn everyone to make good use of them; and note, that if you have the purple mantle of *Sol*, as the sulphur of *Sol*, then be thankful unto God for it, bear no evil mind against your neighbor, unlock your golden seed according to the Tenor of the *Key*, turn it to water; for in our Art there must be Body, Soul, and Spirit, which run together in the innermost root, the one lays hold on the other meliorates the same throughout in its whole quality, insomuch that there is a new created world and earth, which afterwards is illuminated by the Soul, and is exalted into a transcendent efficacy.

Therefore it is requisite that you know, how to infuse your golden seed into the new body, and to bring it to a fluid substance: look about thee, and see where you may find it: if you find none, despair not, but be of good comfort, think upon means, and ask counsel of god *Saturn*, he will not let you go without a resolve, he will put into your hand a deep glittering *Minera* for an offering, which in his Mine is grown of the first matter of all Metals, if this *Minera* after its preparation, which he will show unto you, is set into a strong sublimation, mixed with three parts of bole, or tyle meal, then rises to the highest mount a noble sublimate, like little feathers, or *alumen plumosum*, which in due time dissolves into a strong and effectual water, which brings your seed in a little putrefaction very suddenly into the first volatility, if so be there be added to it a due quantity of water, that it may be dissolved therein, there the twig with the bulk does unite, that they are able to ascend above the highest mountain, and stay inseparably together a Soul and Spirit, or a Spirit and Soul.

It is requisite that you be stored with water for the body or Salt to dissolve the same also, and to coagulate the same into a new clarified body, which will never part asunder, neither in love nor woe, because they are of one nature, nativity, and original, and have been so from the beginning. For they all have their beginning and birth from the power of this volatile bird. But remember well that these Mineral spirits are in other Metals also, and are found effectual in one Mineral, from whence with more ease and less charges it may be had: the business is only herein, that you learn to know, what this Mercurial Spirit, Mercurial Soul, and Astral Salt is, that the one may not be taken instead of the other, which would cause a huge error. You will find, that the nature of the golden Sulphur consists only in all Metals, which are comprehended among the red, and have a fellow dominion with other Minerals, by reason of the fiery tinging Spirits, but the magnetic power and its quality rests in its white Mercurial Spirit, which binds the Soul, and dissolves the body, therefore the *Astrum* of *Sol* is found not only in Gold, that with the addition of the Spirit of Mercury and the *Solar* Salt only the Philosophers stone could be made, but may in like manner be prepared artificially out of Copper and Steel, two immature Metals, both which as male and female have red tinging qualities, as well as Gold itself, whither the same be taken out of one alone, or out of both, being first entered into an Union. Besides, this Mineral in our Mothers tongue is a Mineral, called *Copper* water, and of broken, or digged Verdigreece, or Copper, there can be made a Vitriol, in all which is found gloriously a Soul of the best Gold, and come well to pass very profitably many ways, no Country clown can believe it. Therefore, note here, what you ought to observe, intend thy thoughts, and give not over, unless you come so far, that you know natures mystical conjunction and her dissolution, then you will find, what is requisite for you to know, and return thanks unto thy Creator, make use of it for his glory, and be beneficial to the needy.

This white Spirit is the true Mercury of Philosophers, which has been before me and will be after me, without which the Philosophers stone, and the great mystery cannot be made, neither *universally* nor *particularly*, much less a *particular* transmutation. And this Spirit is the Key to the opening of all Metals, because they have their descent from its sanguinity, as you heard often. For it is that true *primum mobile*, sought of by many thousands, and found by few, and yet all the World is greedy of it, is sought afar off, and found near at hand, it is and moves before the eyes of all men, for if this Spirit be fed with a Metalline Sulphur and Salt, of these three there will be one matter made, not much unlike to the Philosophers great stone, however duly must it be proceeded in, and a true process from the beginning to the end must be observed: for this corporal Salt must be dissolved into this spirit, dissolved, turned, and brought into its *prima materia*, as the spirit himself is: then both these of one equal descent and birth by means of fire with coagulating of the spirit may be generated a third time to a firm fixation, and to a pure transparent white clarified body, then after this accomplished *albedo* the Soul , which is dissolved, must seek for her rest again, penetrate such a pure body, unite with the same, and rise her dwelling therein, that these three be permanent, and abiding constantly in one body eternally clarified.

And that you may be informed, how in this manner both your dissolved seeds, as the spirit of Mercury, and the Soul of the Gold be made again fixed and corporeal, note that it is done only by the proper Salt of *Sol*, which in this Art is called a body. Now observe here, that you take no heterogeneal thing instead of it. What manner of process is here used, read my Fourth *Key*, where the truth of it is held forth with singular examples, and proofs; but you are especially to observe (in case you do not understand that *Key*) this plain and true information, look upon the body of Gold, not as if no other benefit could be reaped of it, but only his Soul, not so: impute no such weakness unto that body, but after you have drawn forth its *Sulphur*, there is yet in it the Salt of glory, and the triumphant victrix, without which your spermatic seed cannot be brought into any coagulation. And even this Salt now, of which I made so long a discourse, how you ought to bring it out of its corporeal form through means of the spirit of Mercury into its *prima materia*, is afterward turned again into a deeply purified and exalted body.

Therefore, take your *Solar* earth, out of which you drew your seeds, or the true *Lions* blood, and reduce it by reverberating to a fixed powder, and subtle impalpable ashes, extract from thence a very subtle Salt, as bright as Ivory is, hereafter I will teach you in the manuals, how the body of *Sol* is anatomized by the *particulars*, and to bring it into a *Sulphur, Salt* and *Mercury*. Then proceed unto the practick and conjunction, and have a care, that you be provident therein, that at their conjunction you do not too much to the one, nor too little to the other, take notice of the quantity, and observe exactly the division of the seeds, hereunto minister a certain measure, and mark my sixth *Key*, then proceed in the begun process, according to the order of the Seventh, Eighth, Ninth, and Tenth *Key's*, as formerly I had informed you about it, go on with it to the appearance of the King's honor and glory, to his highest purple garment, and pure golden piece, is called the triumphing Lord and Conqueror over all his subjects, from the East to the West; which if you have attained unto them return thanks to God, be fervent in praying, be mindful of the poor, be a student unto sobriety, temperance, abstinence, and above all unto taciturnity; for it is the greatest and most heinous sin, to let unworthy men know of it.

The augmentation of this heavenly stone, as also the fermentation is needless to be spoken of in that place, as being described in my last two *Keys*, and held forth to the full, not doubting if God grants so much blessing, and imparts this stone, the sense of these two *Keys* will be more conceived of; for no heterogeneal things must be brought to our Metalline substance, neither at the beginning, middle, or end, but the Mercurial Spirit, and the digested Medicine, spoken of in my eleventh *Key*.

To be further as good as I promised, concerning other things, quoted in my *Keys*, know ye, that no Philosopher is tied wholly unto the Metal of Gold, of which I spoke largely hitherto, and described the true fundamentals thereof, and as you heard afore, the whole mystery lies herein, *viz.* in the tinging of red fiery spirits of Metalline Souls, and all what is tinged red, and is known to have a fierce Sulphur, all such are kind to the *Solar Astrum*, and when the Mercurial spirit is joined with, then the proceedings may go on *Universaliter* and *Particulariter*, that a tincture be obtained from them, whereby Metals and *vulgar Mercury* can be exalted, and be ordered according to the tenor of the process.

Such Souls and goldish *Sulphurs* are found most effectual in *Mars* and *Venus*, as also in *Vitriol*, and both *Venus* and *Mars* can be reduced into a most effectual *Vitriol*, in which Metalline Vitriol afterward all the three *principles*, as *Mercury, Sulphur,* and *Salt* are found under one heaven, and with little pains and short time each can be taken out of it apart, as you shall hear, when I shall make further relation of the Mineral *Vitriol*, which is digged in *Hungary*, of a high gradation. Now if you have wit and understanding, and art inclined and heartily desire to conceive of the true meaning of my *Keys*, and of my other writings, thereby to unlock the locks of Metals for our stone, then you should have taken notice and observed, that in all these I have written not only of the Metal of *Sol*, of its *Sulphur* and *Salt*, but I have interlined and mingled also, and made *uniform*, other red Metals, from whence may be had the *mystical Mastery*: therefore men ought to iterate often the reading of Philosophic books, then a true sense and meaning may be drawn from them, which without divine illumination cannot be neither, *etc.*

But hoping that those, who are fully and really resolved to incline their hearts unto wisdom will give more attention thereunto, than the other mad worldlings, for whom these my writings were not intended: for I spoke as plainly as ever possibly I could, and this kindled light shall further be purified, so that true and sincere Novices may have a full light without an eclipse from their beginning to their ending. To which end I took these pains to disclose that, which all the world was silent in, and concealed it to their last end, and buried it in silence to their very graves.

The scope I aimed at, in so doing, was not to hunt after any vain glory, but rather, that Gods gracious provident goodness might be held forth unto posterity, that the future ages might become seeing, and some of the posterities eyes might be kept open, and be helpful unto their needy fellow Artist, and make them partakers of Gods graces and gifts. Though my mind be mightily perplexed, when I think on what I have done, because I write so plainly, not knowing into whose hands after my departure these my writings may come: However, I hope, let them be what they will , that they will remember, and lay to heart my faithful writings inserted in my former and these present writings, that they deal with this book, and use it so, that they may give good account for it to Almighty God.

Touching further the *Vitriol*, I should make mention of it in my Manuals, where I treat and write generally of Minerals; But it being such a singular Mineral, whose fellow whole nature does not produce, besides, Vitriol before all others is of great affinity unto Metals, and is next kin unto them, for out of all Metals there can be made a Vitriol, or Chrystal (Chrystal and Vitriol is taken for one) therefore I would not bereave it of its own praise, and put its commendation too far off, but rather prefer it, as there is just cause, before other Minerals, and the first place, next to Metals should be given unto it; for (setting aside all Metals and Minerals) this is sufficient to make the Philosophers stone of it, which no other in the world can do the like, though some *particulariter* are a help to further that work, and Antimony alone is a sufficient Master hereunto, as in its due place more shall be spoken of. However, none is thus much dignified in its worthiness, that the said Philosophic stone could be made of it as this Vitriol is. Therefore, ancient Philosophers have concealed this Mineral as much as ever they could, and would not reveal the same to their own Children, that they should not divulge it in the World, but be kept secret, though they published, that such preparation is made out of one thing, and out of one body, which has the nature of *Sol* and *Lune*, and contains also the Mercury, wherein they said true enough, because it is so. But here I must admonish you, that you may turn this argument, and settle your thoughts wholly upon Metalline Vitriols, because I entrusted you, that out of *Venus* and *Mars* there can be made an excellent Vitriol, wherein are found the three principles for the generation of our stone, but you must further note also, that nevertheless these three Metals, as *Spiritus*, *Anima*, *Corpus* are buried and hid in a Mineral Vitriol, as in a Mineral itself. Understand this according to the distinct natures of Vitriol. For the best, which according to my experience showed itself most effectual, herein is that, which is broken, and digged in *Hungary*, of a very deep degree of tincture, not very unlike unto fair blue *Saphir*, having very little of humidities, and other additionals, or strange Ores; the oftener it is dissolved and coagulated, the more is it exalted in its deep tinging color, and is beheld with great admiration.

This high graduated *Vitriol* is found crude in those places, where Gold, Copper, Iron, is broken and digged, and is abundantly transported from thence into foreign parts, insomuch that sometimes there is great scantness of it in those parts, and elsewhere.

Though the vulgar people can afford no better name to it, calling it only a *Copper-Water*, however, ancient Philosophers by reason of its unspeakable virtue and dignity extolled it, and called it *Vitriolum*, for that reason, because its spiritual Oil contains all the three principles of all the triumphing qualities.

If you get such deep graduated and well prepared Mineral , called *Vitriol*, then pray to God for understanding and wisdom for your intention, and after you have calcined it, put it into a well coated Retort, drive it gently at first, then increase the fire, there comes in the form of a white spirit of vitriol in the manner of a horrid fume, or wind, and comes into the Receiver as long as it has any such material in it. And note, that in this wind are invisibly hid all the *three principles*, and come together out of that dwelling, therefore it is not necessary, to seek and search always in precious things, because by this means there is a nearer way open unto nature's mysteries, and is held forth to all such; which are able to conceive of Art and Wisdom.

Now if you separate and free this expelled spirit well and purely *per modum distillationis*, from its earthly humidity, then in the bottom of the glass you will find the treasure, and fundamentals of all the Philosophers, and yet known to few, which is a *red Oil*, as ponderous in weight, as ever any Lead, or Gold may be, as thick as blood, of a burning fiery quality, which is that true fluid Gold of the Philosophers, which nature drove together from the three principles, wherein is found a spirit, soul, and body, and is that *philosophic Gold*, saving one, which is its dissolution, during the fire, and not subject to any corruptibleness, else it flies away with Body and Soul, for neither water nor earth can do it any hurt, because it receives its first birth and beginning from a heavenly water, which in due time is poured down upon the earth.

In these together driven goldish waters lie hid that true bird and *Eagle*, the King with his heavenly *Splendor* together with its clarified *Salt*, which three you find shut up in this one thing and golden property, and from thence you will get all that, which you have need of for your intention.

Therefore set that golden body you have obtained, which in dignity and virtue is exalted beyond all other Gold, into its due and lawful dissolution, its due time, then the Angel of the highest will appear unto you, and tell you that it is the *Sesolver* of all the mysteries in the world, receive it with joy and keep it safe, for its quality is more heavenly than earthly, therefore does it heartily incline to strive after that, which is above, from whence it had its original.

If you have separated this Prophet from his matter which remained, then you need not to undertake any further process, you were taught *parabolically* in my XII. Keys. For even in his remaining formal substance you may find, and expect from thence a pure immortal Soul, together with the glory of the Salt, both which are obtained by means of the spirit, and must be had from thence, and no impure, or contrary thing must be added thereunto. And it is done in the same manner as I told you in my *Keys*, with the Soul and Salt of the Gold by the Saturnal water, in whose place this spiritual Mercurial spirit might be used with better advantage.

Observe only this difference, that the Salt must be drawn forth from the Mercurial body, as it happened unto the Soul, with the spirit of Mercury; whereas on the other side the Salt of Gold must not be drawn forth with the Saturnal-water, because it is too weak for the body of Gold, but with a water, which has been expressed in the description of *particulars*.

This distinction must be exactly observed, being of great concernment, because the Salt of Vitriol is not so strongly guarded, and is not in so fixed a body, as the Gold is, but is still an open body, which saw no coagulation as yet, nor passed it through any melting fire, therefore that body never came as yet to any compactness, there is room left for its own spirit to enter into, can embrace, and unite with its like, and a snow white extraction of Salt may be had, whereas on the other side a sharper matter must penetrate Gold, as you shall hear, when I shall speak more of it in its due place.

Behold now, my friend, whatsoever thou art, what mind I bear towards thee, and how I am affected unto thee in my heart, the like I never durst look for from thee. Consider it well, how sincerely and faithfully I disclose unto you all your locks and bonds, whereby the whole Philosophic wisdom is shut up, which hitherto never entered into any man's thoughts, much less that ever it was practiced, or discovered; and nothing caused me to do it, but only Gods infinite mercy, my good will and love toward my Neighbor, which my Predecessors have not done so completely, and was put off unto me to do it.

Having thus separated your three Mineral bodies, and ordered. them into certain divisions, and put away the dregs, wherein they lie hid, then look to it, that you neglect none of it, by the diminishing of the quantity, which would prove a great fault to your work, and keep each in its own and due quantity, otherwise in your work you cannot come to a happy end.

This is the thrift which so many have missed, and have written great volumes about it: For all that comes from our Philosophic Gold, and has divided itself into three parts, the same must be brought into one, without any loss and diminution, which is to enter into a new form again, and become a meliorated substance, and nothing of it must be done away, but only the *feces terræ* in which the glorious Salt had its dwelling. Therefore, do that I told you of, and join the spirit with the body, bring the body also into a spirit, dissolve and exalt it into the highest spiritual power, in that dissolution the body turns to a spirit, and the spirit with the body unites and joins into one substance, that after the exchanging of all manner of colors, there comes a white body like snow, transcending all whiteness. This is the greatest mystery of this world, about which among the learned and supposed wits, such disputings in the world have been, that a palpable thing, and a visible one could be reduced into its *prima materia*, and out of that may be made again a new clarified and better substance, by the bountiful nature leading the way thereunto.

Thus, you have made and brought into the world the *Queen* of *Honor*, and the *first-born daughter* of Philosophers, which after her due perfection is called the white *Elixir*, of which great volumes are extant. Having brought your work thus far, then you have deserved to be received into the *Turba* of Philosophers, and you get more Art, Wisdom, and Understanding than all Sophisters, which prate much of these mystical things, and yet know not the least thing of it. Therefore, it is just that you should be preferred before them, and let them sit below you in shame and disgrace, and in their darkness of misunderstanding, so long till nature does enlighten then also.

That you nay bring and lead that new Philosophical Creature by the means nature afforded, unto the highest perfection, after which your heart with all her endeavors does strive, then remember that neither man nor beast without a living Soul can neither stir nor move: And as man here in this life, through temporal death loses his Soul, offering the same again to the Almighty God, from whom he had it first, into his mercy and merits of *Jesus Christ*, where after the departure of the Soul the dwelling, as the body of it, is left quite dead, which is buried in the ground, where it rots, and must return unto dust and ashes, being a due reward, which the fall of our first parents in Paradise have deserved, and from then, as by an inheritance is fallen upon us: after which putrefaction there are raised again on the great day new and clarified bodies, and the departed Soul takes her dwelling up again in that new body; after that, there is no more parting of body nor spirit, nor Soul; but because the Soul finding a clarified body, then with the same she makes an everlasting Union, which neither Devil nor Death can destroy, nor disjoin any more, nor bring it into any corruption, but from henceforth into all eternity we are and shall be like unto the best Creatures of God, which before our mortality and departure of the spirit, of the soul and body could never be, God help and grant unto us a blessed resurrection. *Amen.*

This high and mighty example having its foundation, not in humane thoughts, wit, or pride, nor in an ungrounded prating but in the great *Creator's* true word, which he has revealed unto us through his servant and holy prophet *Moses*, does inform you, what you ought to do further with your new begotten Creature, that you may get a perfect birth without any defect, to the praise of the Highest, the Father of lights and mercy, from whom we receive all perfect gifts which he graciously bestowed on his Children, for which we are not able to return sufficient thanks unto him.

Now if you will proceed well in your work, then join the new body with his Soul, which you formerly drew from, that the compound on its virtue be complete, and there be apparent in the end a *plusouam perfection* of it: then is begotten the *Red King* of glory in a fiery substance, and highly clarified body, exalted above all powers upon earth, from thence springs the golden fountain, he that drinks of it is renewed in all his members, and there risen wholly a new life: for the which God be praised forever more.

The augmentation of this huge treasure together with the fermentation thereof, for the transmutation of Metals doubtless you have taken notice of exactly afore, where I wrote of Gold how it must be handled, and what direction I have given you, thereunto, the same you must observe; for here is all one process from the middle to the end, the beginning only asks alteration, by reason of the two distinct matters: for the which God be praised whom we beseech, to give us his grace and blessing, that we may make good use of this treasure, and after this life we may enter into the heavenly Kingdom.

The love of my Neighbor has moved me to write of these things, which in my long experience I found to be true, following the steps of bountiful nature, which made me a Sooth-sayer in natural things, and I am assured, that if these my writings are made public after my death, and my other books sharing in the same fortune, they must undergo many censures. For some will extremely condemn me, delivering me unto Satan, because I have written so plainly: Others there will be, which will quite overthrow my writings, crying them out to be Lies, Superstition, and Diabolical works, the like censure other illuminated men before me have undergone, which they feel to this day; for men are so incredulous in these points, that so mighty an operation should be found against all manner of infirmities, besides the transmutation of Metals in so despicable a matter, (over which the *Iron Man* with his espoused wife *Venus*, together with the deep glittering *Sol*, is, and must have the predominance) and with incredible profit it should by Art be brought to such perfection. The Art being great, and the matter so contemptable, it procures the more doubt and unbelief: those unbelieving men I let understand only this notable example, whereby the eyes of those, that are going unto *Emaus*, shall be opened, and thereby shall acknowledge that I have written no untruth, but disclosed such a truth very plainly. And note, that the ancient Philosophers endeavored to describe the preparation of the Stone under a notion of distilling of wine and the Spirit thereof, which in their work are almost like one to another. For 1. they taught out of the best wine to make a spirit, without any strange phlegm, which to this day among vulgar Artists must be, and is called the right and true mystical spirit of wine, whereas it may soon be proved, that this supposed spirit of wine contains much invisible humidity, or phlegm in an insensible manner, which is nothing else, but its vegetable Mercury: for the fiery spirit of wine is the true fire and soul of the wine. Every Sulphur contains secretly its original and principal Mercury: Vegetables in their kind, the Animals in their kind, and the Minerals also after their kind. 2. They taught how this spirit of wine must be separated in two distinct parts, namely, that this spirit of wine be poured upon white calcined tartar, and be drawn over in a gentle distillation. In this distillation is separated the secret and true spirit of wine from its Vegetable Mercury, as I faithfully informed you in my Manuals. From the remaining earth, they taught a Salt be drawn, to be added to the rectified spirit, whereby it is fortified and strengthened in its substance, and at last the Philosophers stone should be generated. It is mightily against Gods ordinance, that a Vegetable should produce an Animal, or an Animal produce a Mineral. By way of a parable, the practick part is held forth under the notion of this preparation. Now as they taught of the wine, so in like manner also by a short way may our Gold be prepared (not the usual and common Gold) and may be dissolved, divided, separated, and brought into its first principle.

But you must note, that this dissolution and separation was never described plainly by any of the ancient Philosophers which lived before me, and knew the *Magisterium,* why I do it, the love of my Neighbor has moved me thereunto, which I bear from the Center of my heart to all those, which overcame this mystery without falsehood, and shunning vices with a faithful heart, in a sincere knowledge and real piety. In the first place, be informed, that our Gold (so much spoken of hitherto) must never be taken for such Gold by any of our Disciples, which has been melted, and fully digested by nature, for herein such error is committed, that men dilapidate all what they have, and loose both the beginning and end of all their works. Although not only from Gold, but from other Metals also this *Clenodium*, and jewel may be had, in the preparation of it, *particulariter* much profit and advantage may be gotten in that, which concerns men's health, as has been formerly told; however, without the Spirit of Mercury the *Universal* of the World to be gotten merely from the body of *Sol*, is impossible, and will be impossible, unless the Creator of all things produce another ordinance, to change and alter his Creature after his own will. But as that is impossible, so is it impossible also, to deal against Gods Creature in that kind, as to find out that wholesome profit, which to your longing desire you expect. You may believe it for a truth, as Christ himself is, that the Philosophers stone would not be so strange, rare, and unknown a thing, it would be common to Kings and Potentates, if God would permit it to be made of Gold alone, and the three Jewels of infinite fixed virtues hid therein, could be had out of it!

My intent is not in this present Treatise to use any prolixity in writing; those that are not quite blind, and have their eyes open, have enough already to attain unto real knowledge, and command both his mind and hands, not to pass by the weightiest, and esteem high things that are unworthy, and to fall with the blind into the pit, made for them. To those that are real in their desire for to attain unto art and wisdom, and intend to propagate the same without sophistication; and desire in reality to glory in that honorable truth, you may show a real proof of it in this manner.

I tell you really for the highest truth, that you may dissolve *our Gold*, naturally driven together, in a short way, to bring it to its *prima materia*, and is done thus: take the known Mineral Spirit, in which our *Mercury*, *Sulphur* and *Salt* is shut up, containing that Philosophic mystical Gold, pour that *guttatim* upon white calcined tartar, those two contrary qualified matters will be hissing, let them stay together till their contention and strife be ended, and our Gold hide itself invisibly in the vegetable *Salt aere*, or in the belly of tartar: lute a Helmet to it, distill it at first gently in *Balneo*, then increase your fire, then *Hermes* his volatile *bird* will fly away from our Gold in that *Sublimation*, and sit on the highest pinnacles of the Temple, looking about which way to betake himself, but soon is caught in the *Receiver*, which must be pure and very dry; when you see his flight is but slow, then take the glass out of the *Balneo*, set it in ashes, increase your fire, then will she fly more nimbly, keep that fire so long till all is come over, and her brother the *Red Dragon*, hiding his redness under a red color in a whitish fume, will begin to follow after his flying brother. Then cease with the fire, the drops being all fallen from the Helmet, take it off, that which you find in the *Receiver*, you are to keep as a treasure of mysteries. In this manner you have gotten wisdom, understanding, and skill, the fundamentals also, and desires of Philosophers: by this short witty proof you learn and get that knowledge, how this water may be sought after, found, and lighted on, and is not to be esteemed a common water, but is that real infallible *heavenly water*, of which at the beginning I have written, and repeated the same the oftener; which in a spiritual manner from the *heavens power* is poured down upon the earth, begins and accomplishes the generation of all Metals, for that reason the ancient Philosophers called this water *Mercury*, but I call it the *Spirit* of *Mercury*.

Now if you proceed right in this work, and you know what food and what drink to give to this bird, *viz.* Sulphur and Salt of Metal, then you may attain unto the end of the great work, which is almost like unto the Philosophers great work, and you may get profit infinitely *particulariter* many ways; you must note, that this is not the true Philosophic dissolution, but only one, which *particulariter* performs strange matters, and is a *speculum*, in which our Mercury, our *Sol*, and our *Lune* is seen bleaking, which is a present confuting of unbelieving *Thomases*, discovering the blindness of ignorant men. The dissolution of the three *principles* I have described unto you formerly, which is of a slower pace, requiring time and patience, and an exact attention to make, or bring *three* into *one*, which work is done in itself *per se*, without mixing of any heterogeneal matter, only that which lies hid in it, must do it. For the Fountain of Salvation is the illumination of the Soul, and the Salt of the clarified body, are all in that one thing, existing from *one, two,* or *three*, which must be brought and reduced to *one*, which is the golden virtue of all Metals, exalted above all powers, together with the *Eagle* and white body, which are nowhere together, but only in this *one* are found, and in that which is next kin unto it, which knowing Philosophers always held in great esteem, but ignorant and blind men despised and disgraced the same. But those, whose eyes are once opened, love to stick unto much, covet to hide the matter from wicked men, and study day and night how the ignorant might be kept from it. Thus, I close this third part.

And before I begin the fourth part, concerning *Particularia*, I must needs speak something of the Philosophers *Vitriol*, *Sulphur* and *Magnet*.

My friend, you must note, that this description I make now of the essence of Vitriol rests only upon trials made, the victorious triumph of the highest wisdom came by inheritance from the most ancient Philosophers unto me, and comes now unto you, wherein experimentally it's found, that there is a subterranean *Mineral Salt*, called *Vitriol*, which for dyeing of clothes, and many other uses, men cannot well possibly be without it, for it carries on and eats through, by reason of its sharpness, which is distinct from other Salts, in respect of their qualities: for the *Mineral* of this salt is strange, of a very hot and fiery quality, as apparent in its spirit, and contains a twofold spirit, which is *miracutum naturæ*, and is not found the like in other salts and this Salt is an *Hermaphrodite* among other Salts, it is white and red, even as you will have it, it has an extraordinary medicinal quality, performing things in an incredible manner. This Salt contains a combustible *Sulphur*, which is not in other Salts. Therefore, in Metalline affairs touching their transmutation, it performs more than others, because it helps not only to open some, but helps the generation of others, by reason of its *innate* heat. When Vitriol is separated by means of fire, then its spirit at first comes in a white form, after that there comes from its earth a spirit of a red condition, staying in the earth, the Salt being united with its expelled Mercury and Sulphur, can sharpen them: the remainder that stays behind, is a dead earth, of no efficacy. Let this suffice for your learning, and consider well what the Creator holds forth unto you, in nature by this now kindled *ternarie*: for as you find in Vitriols body three distinct things, as Spirit, Oil, and Salt, even so you may expect from its own spirit again (which without the mingling of its Oil, is driven from its matter) three distinct things, even as you did formerly from the body of Vitriol, which deserves very well the name of *Speculum sapientiæ physicæ*, held forth purposely to man to view himself. For if you can separate this Spirit of Vitriol, as it ought, then that affords again unto you three *principles*, out of which only, without any other addition, from the beginning of the world the Philosophers stone has been made: from that you have to expect again a Spirit of a white form, an Oil of red quality, after these two a Christalline Salt, these three being duly joined in their perfection, generate no less than the Philosophers great stone; for that white Spirit is merely the Philosophers Mercury, the red Oil is the Soul, and the Salt is that true Magnetic body, as I told you formerly. As from the Spirit of Vitriol is brought to light the red and white tincture, so from its Oil there is made *Venus* her *tincture*, and in the Center they are much distinct asunder, though they dwell in one body, possessing one lodge; it matters not, for the will of the *Creator* was so, to hide that mystery from unworthy men: observe and consider it well, if so be you intend to be a true follower of Philosophers; In this knowledge lies hid an irrecoverable error, worldly wits cannot conceive of it, that the Spirit of Vitriol, and the remaining Oil should be of so great distinction in their virtue. Touching their properties, the Spirit being well dissolved, and brought into three *principles*, Gold and Silver only can be made by it, and out of its oil only Copper, which will be apparent in a proof made. The condition of the spirit of Vitriol, and its remaining oil is this, that where there is *Copper* and *Iron*, the *Solar* seed commonly is not far from it, and again where there is seed of Gold at hand, Copper and Iron is not far from it, by reason of its attractive Magnetic quality and love, which they, as tinging spirits in a visible manner continually bear one to another. Therefore *Venus* and *Mars* are penetrated and tinged with the super abounding tincture of Gold, and in them there is found much more *root* of the *red tincture*, than in Gold itself, as I made further relation of it in my other books, unto which there belongs also the *Minera* of *Vitriol*, which goes beyond these in many degrees, because its Spirit is mere Gold and *rubedo*, a crude indigested tincture, and in very truth (as God himself is) is indeed not found otherwise.

But this Spirit, as you heard, must be divided into certain distinct parts, as into a spirit, soul, and body, the Spirit is the Philosophic water, which though visibly parted asunder, yet can never be separated *radically*, (because of their unavoidable affinity they bear, and have one to another) as it appears plainly, when afterward they are joined, the one in their mixture embraces the other, even as a Magnet draws Iron, but in a meliorated essence, better than they had before their dissolution. This is the thrift, beginning, middle, and end of the total Philosophic wisdom, affording riches and health, and a long life, it may rather be said, and really proved, that this Spirit is the essence of *Vitriol*, because the Spirit and Oil do differ so much, and were never united radically, because the Oil comes after the spirit, each can be received apart: This fiery spirit may rather and more fitly be called an essence, sulphur, and substance of Gold, and it is so, though it lies lurking in Vitriol as a spirit.

This golden water, or spirit drawn from Vitriol, contains again a sulphur and Magnet, its sulphur is the *anima*, an incombustible fire, the Magnet is its own Salt, which in the conjunction attracts its Sulphur and Mercury, unites with the same, and are inseparable Companions. First in a gentle heat is dissolved the undigested Mercurial spirit, by this is further extracted, after a Magnetic quality, the *sulphureous anima*, in that earth sticks the Salt, which is extracted also in a Magnetic way by the Mercurial spirit, so still the one is a Magnet unto the other, bearing a Magnetic love one to another, as such things were the last together with the *medium* is drawn forth by the first, and are thereby generated, and thus take their beginning. In this separation and dissolution the Spirit, or Mercury is the first Magnet, showing its Magnetic virtue toward the Sulphur and Soul, which it *quasi Magnes* attracts, this spirit *per modum distillationis* being absolved and freed, shows again its Magnetic power toward the Salt, which it attracts from the dead earth; after the spirit is separated from it, then the Salt appears in its purity; if that process be further followed, and after a true order and measure the conjunction be undertaken, and the Spirit and Salt be set together into the Philosophic furnace, then it appears again, how the heavenly spirit strives in a Magnetic way to attract its own Salt, it dissolves the same within XL. days, bringing it to a uniform water with itself, even as the Salt has been before its coagulation. In that destruction and dissolution appears the hugest blackness and *Eclipse*, and darkness of the earth, that ever was seen. But in the exchange thereof a bright glittering whiteness appearing, then the case is altered, and the dissolved fluid waterish Salt turns into a Magnet; for in that dissolution it lays hold on its own spirit, which is the spirit of Mercury, attracts the same powerfully like a Magnet, hiding it under a form of a dry clear body, bringing the same by way of uniting into a deep coagulation and firm fixedness by means of a continued fire, and the certain degrees thereof.

The King with the white Crown being thus generated, and by exsiccation of all humidities being brought to a fixed state, then is it nothing else, but earth and water, though the other Elements be hid therein insensibly; however, both these keep the predominance, though the spirit turn to earth, and can never be seen in a watery form, and this double new born body abides still in its Magnetic quality; for as soon as its departed Soul is restored after its white fixation, then like a Magnet it attracts the same again, unites with it, then are they exalted to their highest tincture and *rubedo*, with a bright transparence and clarity. Thus, in brief you have a short relation of Vitriols, Sulphur, and Magnet. Pray to God for grace, that you may conceive aright of it, put it then to good use, and be mindful of the poor and needy.

At the closing, I annex this briefly, to hold forth unto you a natural proof, that you presently fling and throw down the *Sophister*, and take his Scepter from him. Note, that from all Metals, especially from *Mars* and *Venus*, which are very hard and almost fixed Metals, of each apart can be made a *Vitriol*; this is the reduction of a Metal into a Mineral; for Minerals grow to Metals, and Metals were at first Minerals, and so Minerals are *proxima materia* of Metals, but not *prima*; from these Vitriols may be made, other reductions, namely a Spirit is drawn from them by the virtue of fire.

This spirit being driven over, then there is again a reduction of a Mineral into its spiritual essence, and each spirit in its reduction keeps a Metalline property; but this spirit is not the *prima materia*. Who is now so gross and absurd, that should not be able to conceive further and believe, that by these reductions from one to the other there be a way to *prima materia*, and at last to the seed itself, both of Metals and Minerals: though there be no necessity to destroy Metals, because their seed in the Minerals is found openly fixed.

O good God, what do these ignorant men think, is not this a very easy, and Children's like labor? The one begets the other, and the one comes from the other, is there not bread baked of Corn, upon distinct works? But the World is blind, and will be so to the end of it; Thus much at this time, and commit thee to the protection of the Highest.

End of the Third Part

THE FOURTH PART

OF

BASILIUS VALENTINUS

HIS LAST TESTAMENT

THE MANUALS WHEREIN HE TREATS, HOW METALS AND SOME
MINERALS
MAY PARTICULARITER BE BROUGHT TO THEIR HIGHEST
PREPARATION
& ETC, ETC.

LONDON

Printed by S.G. & B.G. for Edward Brewster,
at the Crane in Saint Pauls Church-yard.
1670

The Fourth Part

Particulars of the Seven Metals, how they may be Prepared with Profit.

First of the Sulphur of Sol
Whereby Lune is tinged into good Gold.

Take of pure Gold, which is three times cast through *Antimony*, and of well purged *Mercury* vive, being pressed through leather, six parts, make of it an *Amalgama*, to the quantity of this Amalgama grind twice as much common Sulphur, let it evaporate on a broad pan in a gentle heat under a muffle, stirring it still well with an Iron-hook, let the fire be moderate, that the matter do not melt together, this Gold calx must be brought to the color of a May-Goldflower, then is it right, then take one part of Saltpeter, one part of Sal Armonic, half a part of grinded pebbles, draw a water from it. Note, this water must be drawn warily and exactly; to draw it after the common way will not do it: he that is used to Chymick preparations, knows what he has to do; and note, you must have a strong stone Retort, which must be coated, to hold the Spirits closely: its upper part must have a pipe, upward of half a spans length, its wideness must bear two fingers breadth, it must be set first in a distilling furnace, which must be open above, that the upper pipe may stand out directly, apply a large receiver, lute it well: let your first fire be gentle, then increase it that the Retort look glowing hot: put a spoonful of this ground matter in at the pipe, close the pipe suddenly with a wet clout, the spirits come rushing into the receiver: these spirits being settled, then carry in another spoonful: in this manner you proceed till you have distilled all. At last give tine to the spirits to be settled, to turn into water: this water is a hellish dissolving strong one, which dissolves instantly prepared Gold calx, and laminated Gold, into a thick solution, of which I made mention above in the third part. This is that water, which I mentioned in my Second *Key*, whi.ch dissolves

146

not only Gold but brings it to volatility, carrying it over the helmet whose *anima* may afterward be drawn from its torn body.

Note, the spirit of common Salt effects the same, if drawn in that manner, which I shall speak of afterward. If three parts of this Salt-spirit be taken, and one part of *spiritus nitri*, it is stronger than Sal armonick water: and is better, because it is not so corrosive, dissolves Gold the sooner, carries it over the helmet, makes it volatile and fit to part with its Soul; you have your choice to use which you think best, and may easier be prepared thus: Take one part of the prepared Gold Calx, and three parts of the water, which you make choice of, put it into a body, lute a helmet to it, set it in warm ashes, let it dissolve, that which is not dissolved, pour three times as much water upon, that all dissolve: let it cool, separate the *feces*, put the solution into a body, lute a helmet to it, let it stand in a gentle heat day and night in *Balneo Mariæ*, if more *feces* be settled, separate them, digest then again in the *Balneo* nine days and nights, then abstract the water gently to *spissitude*, like unto an Oil in the bottom; this abstracted water must be poured on that *spissitude*: this must be iterated often, that it grow weary and weak, remember to lute well at all times. To the oleity on the bottom pour fresh water, which was not yet used, digest day and night firmly closed, then set it in a *sand Capel*, distill the water from it to a thickness: make the abstracted water warm, put it into a body, lute it, abstract it, iterate this work, and make all the Gold come over the helmet.

Note, at the next drawing always the fire must have one degree more: the Gold being come over into the water, abstract the water gently from it in the Balny to the oleity, set the glass into a cold place, there will shoot transparent Crystals, these are the Vitriol of Gold, pour the water from it, distill it again into an oleity, set it by for shooting, more crystals will shoot, iterate it as long as any do shoot. Dissolve these crystals in distilled water, put to it of purged Mercury three times as much, shake it about, many colors will appear, an *Amalgama* falls to the ground, the water clears up, evaporate the *Amalgama* gently under a *muffle*, stirring it still, with a wire, at last you get a purple colored powder, scarlet like, it dissolves in Vinegar into a blood-redness. Extract its *anima* with prepared Spirit of Wine, mixed with the Spirit of common Salt, entered together into a sweetness; this tincture of *Sol* is like a transparent *Rubie*, leaving a white body behind.

Note, that without information you cannot attain unto the Spirit of Salt, if it be not sweet, it has no extractive power; to the attaining hereof, observe these following manuals: take good Spirit of Salt, dephlegmed exactly, driven forth, in that manner, as you shall hear anon.

Take one part of it, add a half a part to it of the best spirit of wine, which must not have any phlegm, but must be a mere Sulphur of wine, and must be prepared in that manner, as I shall tell you anon: lute a helmet to it, draw it over strongly, leave nothing behind; to the abstracted put more spirit of Wine, draw it over, somewhat stronger than you did the first time, weigh it, put a third time more to it, draw it over again, well luted, putrefy this for half a month, or so long as it be sweet, and it is done in Balny very gently: thus the spirit of Wine and Salt is prepared, lost its corrosiveness, and is fit for extracting.

Take the Ruby-red prepared Gold powder, put of this prepared spirit of Salt and wine, so much that it stand two fingers breadth over it, set it in a gentle heat, the spirit will be red tinged, this red Spirit must be canted off, pour new spirit on that which remained on the bottom, set it luted into a gentle heat, let it be tinged deeply, then cant it off, this work must be iterated, that the body of *Sol* remain on the bottom like *calx vive*, which keep, for therein sticks yet more of the Salt of Gold, which is effectual in ways of Medicine, as shall be shown anon.

Those tinged Spirits put together, abstract them gently in *Balneo*, there will be left a red subtle powder in the bottom, which is the true tincture *animated*, or Sulphur of Gold, dulcify it with distilled rain water, it will be very subtle, tender, and fair. Take this extracted Sulphur of *Sol*, as you were taught, and as much of Sulphur of *Mars*, as you shall hear anon, when I treat of *Mars*; grind them together, put it in a pure glass, pour on it so much of Spirit of Mercury, let it stand over it two fingers breadth, that the matter in it may be dissolved, see to it that all dissolve into a Ruby-like Gold-water, jointly drive it over, then is it one, and were at first of one stem, keep it well, that nothing of it evaporate, put it to separated *Silver calx*, being precipitated with pure Salt, and afterwards well edulcorated, and dried, fix it together in a fiery fixation, that it sublime no more: then take it forth and melt it in a wind-oven, let it stream well, then you have united Bride and Bridegroom, and brought them unto Gold of a high degree: Be thankful to God for as long as you live.

I should give further direction, how this extracted Soul of *Sol*, should be further proceeded in, and to make it potable, which ministers great strength, and continued health unto man. But it belonging unto Medicinals, I delay it to that place, where further mention shall be made of.

At this present I will speak only how the white *Solar* body shall further be anatomized, and that by Art its *Mercury vive*, and its *Salt* may be obtained. The process of it is thus:

Take the white body of *Sol*, from which you have drawn its *anima*, reverberate it gently for half an hour, let it become corporeal, then pour on it well rectified hony-water, which is corrosive, extract its Salt in a gentle heat, it is done in ten days space, the Salt being all extracted, abstract the water from it in *Balneo*, edulcorate the Salt with iterated distillings, with common distilled water, clarify it with spirit of wine, then you have *Sol auri*, of which you shall hear more in its due place, of the good qualities it has by way of Medicine upon man. On the remaining matter pour Spirit of *Tartar*, of which in another place, because it belongs unto Medicinals; digest these for a month's time, drive it through a glass Retort into cold water, then you have quick Mercury of *Sol*, many strive to get it, but in vain.

There is one mystery more in Nature, that the white *Solar* body having once lost its *anima*, may be tinged again, and brought to be pure Gold, which mystery is revealed to very few: I shall give a hint of it, that you may not grumble at me to have concealed any point in the work.

I hope you have considered and taken to heart, what I have entrusted you withal about the Universal Stone of Philosophers in my third part, namely how it rests merely upon the white Spirit of Vitriol, and how that all three principles are found only in this Spirit, and how you are to proceed in, and to bring each into its certain state and order.

Take the Philosophic *Sulphur*, which in order is the second principle, and is extracted with the Spirit of *Mercury*, pour it on the white body of the *King*, digest it for a month in a gentle *Balny*, then fix it in ashes, and at last in sand, that the brown powder may appear, then melt it with a fluxing powder made of *Saturn*, then will it be malleable and fair Gold, as it was formerly, in color and virtue nothing defective.

But note, the Salt must not be taken from the *Solar* body, of which I made mention formerly, in a repetition of the XII. Key, where you may read of it. There may be prepared yet in another manner a transparent Vitriol, from Gold in the following manner.

Take good *Aqua Regis* made with *Sal armoniac*, one pound, *id est*, dissolve four ounces of Salmiac in *Aquafort*, then you have a strong *Aqua Regis*, distill and rectify it often over the helmet, let no *feces* stay behind, let all that ascends be transparent. Then take thinly beaten Gold rolls, cast formerly through *Antimony*, put them into a body, pour on it *Aqua Regis*, let it dissolve as much as it will, or as you can dissolve in it: having dissolved all the Gold, pour into some Oil of *Tartar*, or Salt of *Tartar* dissolved in fountain water, till it begins to hiss, having done hissing, then pour in again of the Oil, do it so long that all the dissolved Gold be fallen to the bottom, and nothing more of it precipitate, and the *Aqua Regis* clear up. This being done, then cant off the *Aqua Regis* from the Gold calx, edulcorate it with common water, eight, ten, or twelve times; the Gold calx being well settled, cant off that water, and dry the Gold calx in the air, where the Sun does not shine, do it not over a fire, for as soon as it feels the least heat it kindles, and great damage is done, for it would fly away forcibly, that no man could stay it. This powder being ready also, then take strong Vinegar, pour it on, boil it continually over the fire in a good quantity of Vinegar, still stirring it, that it may not stick unto the bottom, for XXIV. hours together, then the fulminating quality is taken from it: be careful you do not endanger yourself: decant off that Vinegar, dulcify the powder, and dry it. This powder may be driven *per alembicum* without any corrosive, blood-red, transparent and fair, which is strange, and unites willingly with the spirit of wine, and by means of coagulation may be brought to a *Solar* body.

Do not speak much of it to the vulgar: if you receive any benefit by and from my plain and open information, be thus minded, to keep these mysteries secret still to thy dying day, and make no show of it, else thou art naked and lay open to the Devils temptations in all your ways; therefore pray give attention to what I shall tell you, for I will impart unto you this *Arcanum* also, and entrust you upon your conscience with it.

Take good spirit of wine being brought to the highest degree, let fall into it some drops of Spirit of *Tartar*, then take your Gold powder, put to it three times as much of the best and subtlest common flowers of Sulphur, grind these together, set it on a flat pan under a *muffle*, give to it a gentle fire, let the Gold powder be in a glowing heat, put it thus glowing into the spirit of wine, cant off the spirit of wine, dry the powder against a heat, it will be porous. Being dried, then add to it again three parts of *flores sulphuris*, let them evaporate under a *muffle*, neal the remaining powder in a strong heat, and put it in Spirit of wine; iterate the work six times; at last this Gold powder will be so s soft and porous as firm butter, dry it gently, because it melts easily. Then take a coated body, which in its hinder part has a pipe, lute a helmet to it, apply a receiver, set it freely in a strong sand Capel, let your first fire be gentle, then increase it, let the body be almost in a glowing heat, then put in the softened well dried Gold powder, being made warm, behind at the hollow pipe, shoot it in nimbly, there come instantly red drops into the helmet, keep the fire in this degree so long, till nothing more ascends, and no more drops fall into the Receiver. Note, in the Receiver there must be of the best spirit of wine, into which the drops of Gold are to fall.

Then take this spirit of wine, into which the Gold drops did fall, put it in a pelican, seal it *Hermetice*, circulate it for a month, it turns then to a blood-red stone, which melts in the fire like Wax, beat it small, grind among it *Lunar calx*, melt them together in a strong pot, being grown cold, put it in *aquafort*, there precipitates a black *calx*, melt it, then you find much good Gold, as the Gold powder and spirit of wine together with the moiety, and the added *Lunar calx* did weigh, but one moiety of the *Lunar calx* is not tinged, the other is as good as it was to be used. If you hit this rightly, then be thankful to God; if not, do not blame me, I could not make it plainer unto thee.

Now if you will make this Vitriol, then take the powder formerly made, boiled in Vinegar, pour on it good spirit of common Salt, mingled with Saltpeter water, and the spirit of Salt of *Niter*, this Saltpeter water is made, as *aqua Tartari* is made with Saltpeter; Gold is dissolved in this water: which being done, then abstract the water to a thickness, set it in a Cellar, then there shoots a pure Vitriol of *Sol*, the water which stays with the Vitriol must be canted off, distill it again to a *spissitude*, set it in the Cellar, more of the Vitriol will shoot, iterate this work as long as any Vitriol shoots. If you are minded to make the Philosophers Stone out of *Solar Vitriol*, as some fantastic men endeavor in that way, then be first acquainted and ask counsel of thy purse, and prepare ten or twelve pounds of this *Vitriol*, then you may perform the work very well, and the *Hungarian* Vitriol, and others digged out of Mines will permit you to do it. You may extract from this Vitriol also its Sulphur and Salt, with Spirit of wine, which being all easy work, it is needless to describe it.

Now follows the Particular of Lune
And of the Extraction of its Sulphur and Salt.

Take of *calx vive*, and common Salt *ana*, neal them together in a wind-oven, then extract the Salt purely from the *calx* with warm water, coagulate it again, put to it an equal quantity of new calx, neal it, extract the salt from it, iterate it three times, then is the Salt prepared.

Then take the prepared *Lunar calx*, *stratifie* the *calx* with prepared Salt in a glass Viol, pour strong-water on it, made of equal quantities of *Vitriol* and *Saltpeter*, abstract the *aquafort* from it, iterated a third time, at last drive it strongly, let the matter well melt in the glass, then take it forth, your *Lune* is transparent and blueish, like unto an *ultramarine*. Having brought *Lune* thus far, then pour on it strong distilled Vinegar, set it in a warm place, the Vinegar is tinged with a transparent blue, like a *Saphir*, and attracts the tincture of *Lune*, being separated from the Salt, all which comes from *Lune* goes again into the Vinegar, which must be done by *edulcoration*, then you will find the *Sulphur* of *Lune* fair and clear. Take one part of this *Sulphur* of *Lune*, one half part of the extracted *Sulphur* of *Sol*, six parts of the Spirit of *Mercury*, join all these in a body, lute it well , set it in a gentle heat, in digestion, that liquor will turn to a red brown color, having all driven over the helmet, and nothing stand in the bottom, then pour it on the matter remaining of the Silver you drew the Sulphur from, lute it well, set it in ashes for to coagulate, and to fix it XI. days and nights, or when you see the *Lunar* body be quite dry, brown and nothing of it does any more rise, or fume, then melt it quickly with a sudden *flux fire* before the blast, cast it forth, then you transmuted the whole substance of Silver into the best most malleable Gold.

Of this *particular* of Silver, I have made mention in another place, namely in the repetition of my 12. *Keys*, where I wrote that the spirit of Salt also can destroy *Lune*, so that a potable *Lune* can be made of it; Of which *potable Lune* in the last part mention shall be made of. You must note, that further must be proceeded with *Lune*, and a more exact *anatomy* must be made upon *Lune*, thus:

When you perceive that the *Sulphur* of *Lune* is wholly extracted, and the Vinegar takes no more tincture from her, nor the Vinegar does taste anymore of Salt, then dry the remaining *calx* of Silver, put it into a glass, pour on it *corrosive* Hony water, as you did to the Gold, yet it must be clear, and without any *feces*, set it in a warmth for four, or five days, extract *Lune's Salt*, which you may perceive, when the water grows white. The Salt being all out of it, then abstract the Hony water, *edulcorate* the corrosiveness by distilling, and clarify the Salt with spirit of wine, the remaining matter must be *edulcorated* and dried, pour upon it spirit of *Tartar*, digest it for half a month, then proceed as you did with the Gold, then you have *Mercury* of *Lune*. The said Salt of *Lune* has excellent virtues upon man's body, of which I shall speak in another place. The efficacy of its Salt and Sulphur, may be learned by this following process.

Take of the sky-colored Sulphur, which you extracted from *Lune*, and is rectified with spirit of wine, put it in a glass, pour on it twice as much of spirit of *Mercury*, which is made of the white spirit of *Vitriol*, as you have heard in the same place. In like manner take of the extracted and clarified Salt of Silver, put to it three times as much of spirit of *Mercury*, lute well both glasses, set then into a gentle Balny for eight days and nights, look to it that the Sulphur and Salt lose nothing, but keep their quantity as they were driven out of the Silver. Having stood these eight days and nights, then put them together into a glass, seal it *Hermitice*, set it in gentle ashes, let all be dissolved, and let it be brought again into a clear and white *coagulation*, at last fix them by the degrees of fire, then the matter will be as white as snow, thus you have the white tincture, which with the *volatile* dissolved *anima* of *Sol* you may animate, fix, bring to the deepest redness, and at last *ferment*, and *augment* the same in *infinitum*, the spirit of *Mercury* being added thereunto. And note, that upon Gold a process is to be ordered, with its *Sulphur* and *Salt*.

If you understand how their *primum mobile* is to be known, then is it needless in this manner, and to that purpose to destroy Metals, but you may prepare everything from, or of their first essence, and bring them to their full perfection.

Of the Particular of Mars
Together with the Extraction of its Anima and Salt.

Take of red Vitriol Oil, or Oil of Sulphur one part and two parts of ordinary Well-water, put those together, dissolve therein filings of Steel, this dissolution must be filtered being warmed, let it gently evaporate a third part of it, then set the glass in a cool place, there will shoot Crystals as sweet as Sugar, which is the true *Vitriol* of *Mars*, cant off that water, let it evaporate more, set it again in a cold place, more Crystals will shoot, neal them gently under a muffle, stirring still with an Iron wire, then you get a fair purple colored powder, on this powder cast distilled Vinegar, extract the *anima* of *Mars* in a gentle Balny, abstract again the Vinegar, and *edulcorate* the *anima*. This is the *anima* of *Mars*, which being added to the spirit of *Mercury*, and united with the *anima* of *Sol*, tinges *Lune* into *Sol*, as you heard about the Gold.

Of the Particular of Venus
What mysteries there are hid therein,
and of the Extraction of its Sulphur and Salt.

Take as much of *Venus* as you will, and make *Vitriol* of it, after the usual and common practice, or take good Verdigreece, sold in shops, it effects the same, grind it small, pour on it good distilled Vinegar, set it in a warmth, the Vinegar will be transparent green, cant it off, pour on the remaining matter on the bottom new Vinegar, iterate this work as long as the Vinegar takes on any tincture, and the matter of the Verdigreece on the bottom lies very black: put the tinged Vinegar together, distill the Vinegar from it to a dryness, else a black Vitriol will shoot, thus you get a purified Verdigreece, grind it small, pour on it the juice of immature Grapes, let it stand in a gentle heat, this juice makes a transparent tincture as green as a *Smarag'd,* and attracts the red tincture of *Venus*, which affords an excellent color for Painters, Limmers, and others for their several uses.

When the juice extracts no more of the tincture, then put all the extractions together, abstract the moiety of this juice gently, set it into a cool place, there shoots a very fair Vitriol, if you have enough of that, then you have matter enough, to *reduce* the same, and to make of it the Philosophers stone, in case you should make a doubt to perform this great mystery by any other Vitriol. Of this preparation I have spoken already *Parabolicè* in the book of the *Keys*, in the Chapter of the Wine-Vinegar, where I said, that the common *Azoth* is not the matter of our stone, but our *Azoth*, or *materia prima* is extracted with the common *Azoth*, and with the Wine, which is the out-pressed juice of unripe Grapes, and with other waters also must be prepared, these are the waters wherewith the body of *Venus* must be broken, and be made into Vitriol, which you must observe very well, then you may free yourselves from many troubles and perplexities.

But especially note, that the way of the *Universal* with this vitriol is understood in the same manner, and is thus conditioned, as I told you in the third part of the *Universal,* and pointed at the common *Hungarian Vitriol,* and even as well out of *Mars,* put *Particulariter* to be dealt upon with *Venus.* Therefore know, that it may be done with great profit, if you drive forth the red oil of Vitriol, and dissolve *Mars* in it. And Crystallize the solution, as you were told, when I treated of *Mars.* For this dissolution and coagulation *Venus* and *Mars* are united, this Vitriol must be nealed under a *muffle* into a pure red powder, and must be extracted further with distilled Vinegar, as long as there is any redness in it, then you get the *anima* of *Mars,* and of *Venus* doubled, of this doubled virtue after the addition of the *anima* of *Sol,* which you made in the before quoted quantity take twice as much of *Silver Calx*, and fix it, as you heard when I spoke of the *Particular* of *MARS* and *Sol*.

But note, that there must be twice as much of the spirit of *Mercury*, then there was allowed in that place, but in the rest the process is alike. The Salt of *Venus* must be extracted when the juice takes no more of the green tincture, then take the remaining matter, dry it, pour Honey[15] water upon it, then that Salt goes in that heat for five, or six days, and clarify it with spirit of Wine, then is the Salt ready for your Medicine.

Of the Particular of Saturn
Together with the Extraction of its Soul and Salt.

Most men hold and count *Saturn* an unworthy and mean Metal, and is abused most basely in several expeditions, whereas, if known in its internals, more laudable exploits would be performed with it, and many excellent Medicines be prepared of it. Being it is my intention to put an Elucidation to my former writings, to leave it after me for a *Legacy* unto posterity, that simple men of ordinary capacity might know and conceive also of the things I formerly wrote of, which after the resurrection of my flesh myself shall bear record unto, that I have written more than was meet, which others before me have purposely concealed; it being my purpose to declare fundamentally all such *Particulars*, which formerly at large I discoursed of in a Philosophic manner, thus; that this my Declaration made in my decrepit age be noted conscionably by those into whose hands it comes, that this my Revelation, which in Gods providence will be disposed of, to be a lamp of truth unto all the world, may not be imparted unto men unworthy of Gods mysteries, which acknowledge not the Creator of them in a pure and humble and penitent heart, persevering conversation, and a fervent purpose to incline towards him. This present writing I leave as a precious badge with an earnest proviso, that men would look and observe carefully every letter contained in this, and other of my writings, which in all fidelity I hold forth unto them: And begin now with *Saturn*, who in all probability after Astronomic rules is the highest and chiefest Lord in the celestial spheres, by whose influence the *subterranean Saturn* has its life and coagulation, putting that black color on it, the rest from the best to the worst follow after, whose splendor enlightens that whole firmament, and is incorruptible.

I should speak something of *Saturn's Nativity*, from whence he takes his off-spring, but in this place, I do not hold it requisite (being there has been mention of it in several places in my other books) because it is to no purpose for novices, and to

---

[15] Interesting to note: here the word is spelled with an "e": Honey instead of "Hony" -pnw

repeat all, would increase the volume, which I do not intend, purposing only to elucidate such things, which formerly have been delivered in obscure terms.

Note, *Saturn* is not to be thus slighted by reason of its external despicable form, if he be wrought in a due process after the Philosophers way, he is able to requite all the pains the Art seeking Laborers bestow on him, and will acknowledge him rather to be the Lord, and not the Servant: A Lords honor is due unto him, not only in respect of man's health, but in respect also of meliorating of Metals: the preparation of it is thus:

Take red Minium, or Ceruse, these are of several worths, the one is better before the other, according to their several examinations, those that are sold in shops are seldom pure, without their due additionals: my advice is, that every Artist undertake himself the destruction of *Saturn*, the process of it is several, of the best I give this hint:

Take pure Lead, which yields to the hammer, as much as you please, laminate it thinly, the thinner the better, hang these lamins, in a large glass filled with strong Vinegar, in which is dissolved a like quantity of the best Sal Armoniac, sublimed thrice with common Salt, stop the glasses mouth very closely, that nothing evaporate, set the glass in ashes of a gentle heat, otherwise the spirits of the Vinegar and Sal Armoniac ascend, and touch the Saturnal lamins, at the tenth, or twelfth day you will spy a subtle Ceruse hanging on these lamins, brush them off with a Hares foot, go on, get enough of this Ceruse, provided, you buy good wares, if sophisticated, you labor in vain. Take a quantity of it, if you please, put it in a body, pour strong Vinegar on it, which several times has been rectified, and was fortified at the last rectification with a sixteenth part of spirit of vulgar Salt, dephlegmed, and drawn over: stop the body well, or which is better, lute a blind-head to it, set the body in ashes to be digested, swing it often about, in a few days the Vinegar begins to look yellow and sweet, at the first, iterate it a third time, it is sufficient. The remnant of the Ceruse stays in the *bodies* bottom unshapely, filter the tinged Vinegar clearly, that is of a transparent yellowness: put all the tinged Vinegar together, abstract two parts of it in *Balneo Mariæ* let the third part stay behind, this third part is of a reasonable *Rubedo*, set the glass in a very cold water then the Crystals will shoot the sooner, being shot, take them out with a wooden spoon, lay them on a paper for to dry, these are as sweet as Sugar, and are of great energy against inflamed symptoms: abstract the Vinegar further in *Balneo*, in which the Crystals did shoot, set that distillation aside, for the shooting of more Crystals, and proceed with these as you did formerly.

Now take all these Crystals together, they in their appearance are like unto clarified Sugar, or Saltpeter, beat them in a Mortar of Glass, or Iron, or grind them on a Marble unto an unpalpableness, reverberate it in a gentle heat, to a blood-like redness: Provided, they do not turn to a blackness. Having them in a Scarlet color, put them in a glass, pour on a good spirit of Juniper, abstracted from its oil, and rectified several times into a fair, white, bright manner, lute the glass above, set it in a gentle heat, let the spirit of Juniper be tinged with a transparent redness like blood, then cant off neatly from the *feces* into a pure glass: with that proviso, that no impure thing run therewith, on the *feces* pour other Spirit of Juniper, extract still, as long as any spirit takes the tincture: keep these *feces*, they contain Salt.

Take all these tinged spirits together, filter them, abstract them gently in *Balneo*, there remains in the bottom a neat Carnation powder, which is the *anima* of *Saturn*, pour on it Rainwater, often distilled, distill it strongly several times, to get off that, which stayed with the spirit of Juniper, and so this subtle powder will be edulcorated delicately: keep it in a strong boiling, cant it off, then let it go off neatly, let it dry gently, for safety's sake, reverberate it again gently for its better exsiccation, let all impurity evaporate, let it grow cold, put it in a Viol, put twice as much of spirit of Mercury to it, which I told you of in the third part of the *Universal*, entrusted you upon your conscience with it, seal it *Hermeticè*, set it in a vaporous Bath, which I prescribed at the preparation of the spirit of Mercury, called the Philosophers *fimus equinus*, let it stand in the Mystical Furnace for a month, then the *anima* of *Saturn* closes daily with the spirit of Mercury, and both become inseparable, making up a fair transparent deeply tinged red Oil; look to the government of the fire, be not too high with it, else you put the spirit of Mercury as a volatile spirit to betake himself to his wings, forcing him to the breaking of the glass: but if these be well united, then no such fear look for, for one nature embraces and upholds the other.

Then take this Oil, or dissolved *anima* of *Saturn* out of the Viol, it is of a gallant fragrancy, put it into a body, apply a Helmet to it, lute it well, drive it over, then Soul and Spirit is united together, and fit to transmute Mercury precipitated. into *Sol*.

The precipitation of Mercury is done thus: take one part of the spirit of Salt of Niter, and three parts of Oil of Vitriol, put these together, cast into it half a part of quick Mercury, being very well purged, set it in Sand, put a reasonable strong fire to it, so that the spirits may not fly away, let it stand a whole day and night, then abstract all the spirits, then you find in the bottom a precipitated Mercury, somewhat red, pour the spirits on again, let it stand day and night, abstract it again, then your precipitate will be more red than at first, pour it a third time upon it, then abstract strongly, then your *precipitate* is at the highest *rubedo*, dulcify it with distilled water, let it strongly be exsiccated. Then take two parts of this precipitated Mercury, one part of the dissolved *Saturnal* Oil, put these together, set it in ashes, let all be fixed, not one drop must stick anywhere to the glass. Then it must be melted with due additionals of lead; they close together, afford Gold, which afterward at the casting through *Antimony* may be exalted.

I have informed you hereof where I treated of *Mercury vive*. But note, that Mercury must not be precipitated, unless with pure Oil of *Vitriol*, or Oil of *Venus*, with the addition of the spirit of Salt Niter: Albeit such Mercury cannot be brought to its highest fixation, by way of precipitating, but its fixed coagulation is found in *Saturn*, as you heard.

Beat the above said Mercury small , grind it on a stone, put it in a Viol, pour on it the distilled *Saturnal* Oil, it enters instantly, if so be you proceeded right in the precipitation, seal the Viol, *Hermeticè*, fix it in ashes, at last in sand, to its highest fixation, then you have bound Mercury with a true knot, and brought him into a fixed coagulation, which brought its form and substance into a melioration, with an abundance of riches, if you carry it on a white precipitate, then you get only silver, which holds but little of Gold.

One thing I must tell you about this process, that there is yet a better way to deal upon *Saturn*, with more profit, that you may not have any cause to complain against my not declaring it, take it thus: take two parts of the above said dissolved Oil, or of the *Saturnal* Soul, one part of *Astrum Solis*, and of *Antimonial Sulphur*, whose preparation follows afterwards, two parts, half as much of *Salt* of *Mars*, as all these are, weigh them together, put all into a glass Vial, let the third part of it be empty, set it in together to be fixed, then the Salt of *Mars* opens in this compound, is fermented by it, and the matter begins to incline to a blackness, for ten, or twelve days it is eclipsed, then the Salt returns to its coagulation, laying hold in its operation on the whole compound, coagulate it first into a deep brown mass, let it stand thus unstirred in a continued heat, it turns to a blood red body, increase the fire, that you may see the *Astrum Solis* to be predominant, which appears in a greenish color, like unto a Rainbow: keep this fire continually, let all these colors vanish, it turns to a transparent red stone very ponderous, needless to be projected on Mercury, but tinges after its perfection, and fixation all white Metals into the purest Gold. Then take of the prepared fixed red stone, or of the powder one part, and four parts of any of the white Metal, first let the Metal melt half an hour, and let it be well clarified, then project the powder upon it let it drive well, and see that it be entered into the Metal , and the Metal begin to congeal, then is it transmuted into Gold, beat the pot in pieces, take it out, if it has any Slacks, drive them with Saturn, then is it pure and malleable. If you carry it on *Lune*, then put more of the powder to it then you do upon *Jupiter* and *Saturn*, as half an ounce of the powder tinges five ounces of *Lune* into *Sol*, let this be a miracle, fool not thy Soul with imparting this mystery unto others, that are unworthy of it. Proceed with *Salt* of *Saturn*, as you were informed about *Mars* and *Venus*, only distilled Vinegar performs that, which Honey water did by the other, and clarify it with spirit of Wine.

Of the Particular of Jupiter
Together with the Extraction of its anima and Salt.

Take Pumice-stones, sold in shops, neal them, quench them in old good Wine, neal them again, and quench them as you did formerly, let this nealing be iterated a third time, the stronger the Wine is you quench withal, the better it is, after that dry them gently, thus are they prepared for that purpose. Pulverize these Pumice-stones subtly, then take good Tin, laminate it, stratify it in a cementing way in a reverberating furnace, reverberate this matter for five days and nights in a flaming fire, it draws the tincture of the Metal, then grind it small, first scraping the Tin-lamins, put it in a glass body, pour on it good distilled Vinegar, set it in digestion, the Vinegar draws the tincture, which is red-yellow, abstract this Vinegar in *Balneo*, edulcorate the *anima* of *Jupiter* with distilled water, exsiccate gently, proceed in the rest as you did with the *anima* of *Saturn*, *viz.* dissolve radically in, or with the spirit of *Mercury*, drive them over, pour that upon two parts of red *Mercury* precipitated, being precipitated with this *Venerean* sanguine quality, then coagulate and fix: if done successfully, you may acknowledge *Jupiter's* bounty, that gave leave to transmute this *precipitate* into Gold which will be apparent at their melting. It performs this also, it transmutes ten parts of *Lune* into *Gold*, if other *Sulphurs* be added thereunto: force no more upon *Jupiter*, it's all he is able to do, being of a peaceable disposition, he told all what he could do. The process about this Salt, is, to extract it with distilled rain-water, clarified with spirit of Wine.

Of the Particular of Mercury vive
And of its Sulphur and Salt.

Take of quick *Mercury*, sublimed seven times, *lib. semis*, grind it very small, pour on it a good quantity of sharp Vinegar, boil it on the fire for an hour, or upward, stirring the matter with a wooden spatula, take it from the fire, let it be cold, the *Mercury* settles to the bottom, and the Vinegar clears up: if it be slow in the clearing, let some drops of Spirit of *Vitriol* fall in the Vinegar, it does precipitate the other, for *Vitriol*, precipitates *Mercury vive*, *Salt* of *Tartar* precipitates *Sol*, *Venus* and *common Salt*, does precipitate *Lune*, and *Mars* does the like to *Venus*, a *lixivium* of Beech-ashes does it to *Vitriol*, and *Vinegar* is for *common Sulphur*, and *Mars* for *Tartar*, and *Saltpeter* for *Antimony*. Decant off the Vinegar from the *precipitate*, you will find the *Mercury* like a pure washed Sand, pour on it Vinegar, iterate this work a third time, then edulcorate the matter, let it dry gently.

Take two ounces of *anima* of *Mars*, one ounce of *anima* of *Saturn*, one ounce of *anima* of *Jupiter*, dissolve these in six ounces of *Mercurial* spirit, let all be dissolved, then drive it over, leave nothing behind, it will be a Golden water, like a transparent dissolution of *Sol*, your prepared and edulcorated *Mercury* must be warmed in a strong Viol, pour this warmed water gently on it, a hissing will be, stop the Viol, then the hissing is gone; then seal it *Hermeticè*, set it in a gentle Balny, in ten days the *Mercury* is dissolved into a grass green Oil, set the Viol in ashes for a day and night, rule your fire gently, this green color turns into a yellow oil, in this color is hid the *Rubedo*, keep it in the fire, and let the matter turn to a yellow powder, like unto Orpiment; when no more comes over, then set the glass in sand for a day and a night, give a strong fire to it, let the fairest *Ruby-rubedo* appear, melt it to a fixedness with a fluxing powder made of *Saturn*, it comes now to a malleableness, one pound of it contains two ounces of good Gold, as deep, as ever Nature produced any. Remember the poor, do not precipitate thyself into an infernal abyss, by forgetting thyself in not doing the duties you ought to perform in regard of the blessing.

An Oil made of Mercury, and its Salt.

Take quick Mercury, being often sublimed, and rectified with *Calx vive*, put it in a body, dissolve it in a heat, in strong *Nitrous* water, abstract the water from it, the corrosiveness which stays there, must be extracted with good Vinegar, well boiled in it: at last abstract this Vinegar, the remainder of it must be dulcified with distilled water, and then exsiccated. Afterward on each pound must be poured *lib.* 1. of the best spirit of Wine, let it stand luted in putrefaction, then drive over what may be driven, first gently, then more strongly, from that which is come over, abstract the spirit of Wine *per Balneum*, there stays behind a fragrant Oil, which is *Astrum Mercurii*, an excellent remedy against Venereal diseases.

Seeing the *Salt* and *Astrum* of *Mercury* is of the same Medicinal operation, I hold it needless to write of each in particular, and will join their operation into one, and declare of it in the last part about the Salt of Mercury, because they are of one effect in Medicinal operations. Take the made Oil, or *Astrum Mercurii*, which by reason of its great heat keeps its own body in a perpetual running, casting it on the next standing earth, from which you formerly drew the Oil. Set it in a heat, the Oil draws its own Salt; that being done, put to it a reasonable quantity of spirit of Wine, abstract it again, the Salt stays behind, dissolved in the fresh spirit of Wine, being *dulcified* by *cohobation*. Then is the *Mercurial Salt* ready, and prepared for the Medicine, as shall be mentioned in the last part. *Mercury* is able to do no more, neither *Particulariter*, nor *Universaliter*, because he is far off from *Philosophers Mercury*, although many are deceived in their fancies to the contrary.

Of the Particular of Antimony
Together with the Extraction of its Sulphur and Salt.

Take good *Hungarian* Antimony, pulverize it subtly to a meal, calcine it over a gentle heat, stirring it still with an Iron wire, and let it be *albified*, and that at last it may be able to hold out in a strong fire. Then put it into a melting pot, melt it, cast it forth, turn it to a transparent glass, beat that glass, grind it subtly, put it in a glass body of a broad flat bottom, pour on it distilled Vinegar, let it stand luted in a gentle heat for a good while, the Vinegar extracts the Antimonial tincture, which is of a deep redness, abstract the Vinegar, there remains a sweet yellow subtle powder, which must be edulcorated with distilled water, all acidity must be taken off, exsiccate it; pour on it the best graduated spirit of Wine, set it in a gentle heat, you have a new extraction, which is fair and yellow, cant it off, pour on other spirit, let it extract as long as it can, then abstract the spirit of Wine, exsiccate, you find a tender deep yellow subtle powder of an admirable Medicinal operation, is nothing inferior unto potable *Sol*.

Take two parts of this powder, one part of *Solar Sulphur*, grind these small, then take three parts of *Sulphur* of *Mars*, pour on it six parts of *Spirit* of *Mercury*, set it in digestion well luted, let the *Sulphur* of *Mars* be dissolved totally, then carry in a fourth part of the ground-matter of the Sulphur of *Antimony*, and of *Sol*, lute and digest, let all be dissolved, then carry in more of your ground Sulphurs, proceed as formerly, iterating it so long till all be dissolved, then the matter becomes a thick brown Oil, drive all over jointly into one, leave nothing behind in the bottom, then pour it on a purely separated *Lunar calx*, fix it by degrees of fire, then melt it into a body, separate it with an *Aquafort*, six times as much of *Sol* is precipitated then, above the ponderosity the compound did weigh, the remainder of *Lune* serves for such works you please to put it unto.

The *Antimonial tincture* being extracted totally from its *Vitrum*, and no Vinegar takes more hold of any tincture, then exsiccate the remaining powder, which is of a black color, put it into a melting pot, lute it, let it stand in a reasonable heat, let all the sulphureous part burn away, grind the remaining matter, pour on it new distilled Vinegar, extract its Salt, abstract the Vinegar, edulcorate the *acidity* by *cohobation*, clarify so long, so that the water be white and clear. If you have proceeded well in your manuals, then the lesser time will be required to extract the *Antimonial Salt*, as you shall hear of it. Whereby you may observe, that the *Antimonial Sulphur* is extracted in the following manner, and is of the same Medicinal operation, but is of a quicker and speedier work, which is a matter of consequence, and worthy to be taken notice of.

A Short way to make Antimonial Sulphur and Salt.

Take good Vitriol, common Salt, and unslaked lime, of each one pound, four ounces of Salt-armoniac, beat them small, put them in a glass body, pour on it three pounds of common Vinegar, let it stand in digestion stopped for a day, put it afterward into a Retort, apply a receiver to it, distill it, as usually an *aquafort* is distilled. Take of the off drawn liquor, and of common Salt, one pound each, rectify them once more, let no muddiness come over with it, all must come clear: then take one pound of pulverized antimonial glass, pour this spirit on it, lute well, digest, and let all be dissolved; then abstract the water in *Balneo Mariæ*, there remains in the bottom a black, thick, fluid matter, but somewhat dry, lay it on a glass Table, set it in a Cellar, a red Oil flows from it, leaving some *feces* behind, coagulate this red Oil gently upon ashes, let it be exsiccated there; then pour the best spirit of Wine on it, it extracts a tincture which is blood red, cant off that which it tinged, pour other spirit of Wine on the remainder, let all redness be extracted, thus you have the tincture or *Antimonial Sulphur*, which is of a wonderful Medicinal efficacy, and is *equivalent* unto potable Gold, as you heard in the former process. And its preparation serves now to proceed with its *Particulariter*, as I showed in the former. This black matter, which stayed behind after the extraction of *Sulphur*, must be well exsiccated, extract its snow-white Salt with distilled Vinegar, edulcorate it, clarify it with spirit of Wine, observe its virtues in *Medicina*, of the which in the last part.

Thus, I conclude my fourth part also. Other mysteries in Nature, and some augmentations might be here annexed, but I wave them, mentioning only the chiefest of them, and are such, which may be wrought easily, and in a short time, and whereby good store of riches may be gotten. The rest, which are not of that importance, and nay easily draw *Novices* into errors, bringing no profit for the present, may in good time by careful practice be found out and obtained.

If you only know these, whereby health and wealth is obtained, then these metalline Sulphurs in their compounds may bring great profit unto you, to write of all these circumstantially, is impossible to one man, it is of an infinite labor; call upon God for grace and mercy: A fundamental Theory affords the practick part, from thence flow infinite springs, all from one hand. If you go otherwise to work, then I entreated you to do by the Creator of heaven and earth, then all your actions will be retrograde unto a temporal disaster.

I should annex here the efficacies of other Minerals, which are next unto Metals: but seeing they are of no ability unto transmutation of Metals, but are only Medicinal, and are qualified to do their work to the admiration of those that make use of them, I leave them at this time. The Almighty has put wonderful virtues into Metalline Salts, which have been found approved several ways.

END OF THE FOURTH PART.

A PRACTICK TREATISE

TOGETHER WITH THE

XII. KEYS[16]

AND

APPENDIX

OF THE

GREAT STONE

OF THE

ANCIENT PHILOSOPHERS.

Written and left by *Basilius Valentinus*
a *German* Monk of the Order of St. *Bennet*.

LONDON
Printed by S. G. & B. G. for Edward Brewster,
at the Crane in
Saint Pauls Church-yard, 1670.

---

[16] This is *Twelve Keys of Basilius Valentinus*, similar to the version that appears in The R.A.M.S. Library of Alchemy Vol. 1. Comparing the two might prove useful. -pnw

The Preface of the Author

Being possessed with human fear, I began to consider, out of the simplicity of Nature, the miseries of this World, and exceedingly lamented with myself the offences committed by our First Parents, and how little repentance there was throughout the world, and that men grew daily worse and worse, an eternal punishment without redemption hanging over the heads of such impenitents: Therefore made I haste to withdraw myself from sin, and bid farewell to the World, and addict myself to the Lord as his only Servant.

Having lived some time in my Order, then also, after I had done my appointed devotions, meddling not with frivolous things, least my vain thoughts through idleness should yield causes of greater evils; I took upon me diligently to search into Nature, and thoroughly to Anatomize the *Arcanums* thereof, which f found to be the greatest pleasure next to Eternal things. Having found in our Monastery many books written by Philosophers of ancient time, who had truly followed Nature in their study and search; this gave a greater encouragement to my mind, to learn those things they knew; and though it proved difficult to me in the beginning, yet at last it proved more easy. The Lord so granted (to whom I daily prayed) that I should see those things that others before me had seen.

In our Monastery, one of my Fellows was much tormented with the Stone, that he often tiles lay bed rid, had sought many Physicians, and despaired of any help from them, resigned up his life to God, having laid aside all human help.

Then began I to Anatomize vegetables, and distilled them, I extracted their Salts and Quintessence: But amongst all these, could I not find anything, that would free my sick brother from his distemper, although I tried many things, for they were not so effectual in their degrees to cure that disease, so that for six years' space, there was hardly any vegetable, that I had not in some way or other wrought upon.

Then I bent my thoughts to consider further of this matter, and to addict myself to a fundamental knowledge, and search after these hidden virtues which the Creator had placed in Metals and Minerals. The more I sought into them, the more I found, one Secret still flowing from another; God prospered my endeavors, that I tried many things, and my eyes also saw these virtues, which Nature had infused into Metals and Minerals, yea, various, that they are not easily understood by the ignorant and slothful.

Amongst all these I happened on a certain Mineral, composed of many colors, and of very great power in Art, I extracted its spiritual Essences and thereby in a few days I restored my sick brother to his former health: For this spirit was so strong, that it did much revive or fortify the spirit of my Brother, who as long as he lived daily prayed for me for he lived long after, and then bid me farewell. His and my prayers do so much prevail, that the Creator, discovered, and by reason of my diligence did demonstrate unto me, even that, which yet remains hid to the wise men, as they call themselves.

So therefore, in this treatise will I declare, and so far as is lawful for me to do, reveal, The Stone of the Ancients granted unto us for the health and comfort of man in this Valley of misery, as the chiefest of all Earthly Treasure. Writing these things not for my own but the benefit of posterity, following therein the method I find in the writings of many very learned men, so that by my writings, the dictates of Philosophy which are very short and Enigmatical, you may attain that Rock on which Truth depends, with a temporal reward and eternal blessings. AMEN.

Of the great Stone of the ancient Philosophers.

Dear Friend and lover of Art, in my Preface I promised to show unto thee, and to such others who are very desirous to learn the Properties of Nature, and diligent searchers into Art, that Corner Stone, and that Rock, so far as I am permitted from above, as our Ancestors the Ancients prepared their Stone, which they attained from the Most High, for the preservation of their health, and for their benefit in this present world. That I may therefore perform my promise, and not lead you into Labyrinths by Sophistic errors, I will reveal unto you the Fountain of all good things; therefore, observe my following words, and diligently weigh them, if you have a desire to learn this Art: I shall not use much Eloquence, that is not my intent, very little will be learned from that, I delight in brevity, which shall contain the fountain of the matter.

Know, that very few have attained unto the possession of this Magistery, although many have labored and wrought in our Stone, but the true knowledge and obtaining thereof, the Creator has not made common, but will grant the same to such as are averse to lies, and love the truth, and which with humble hearts most diligently seek the said Art, especially to such who love God unfeignedly, and pray unto him therefore.

Wherefore I tell thee for a truth, if you would make our great and ancient Stone, follow my Doctrine, and above all things pray to the Maker of every Creature, that he may bestow on you his grace and blessing to that end; and if you have sinned, confess and deal righteously, and resolve upon it, that you sin no more, but live holily, that your heart may be filled with every good thing; and remember when you are preferred in honors, to be helpful to the poor and indigent, that you deliver them from their miseries, and refresh them with your bountiful hand, that you may obtain the greater blessing from the Lord, and through the Confirmation of faith receive your Throne in Heaven prepared for you.

My friend despise not, nor contemn the real writings of such men, who had the Stone before us; for next unto Revelation of God I obtained it from them, and let the reading of them be many times and very often reiterated, least you forget the foundation, and the truth be extinguished as a Lamp.

Then be not unmindful of your diligent labor, always seeking in the writings of Authors; and be not of an unstable mind, but rely on that fixed Rock, wherein all wise men do unanimously concur, for a wavering man is soon led into a wrong way, and precipitates himself into many errors; and men of wavering minds seldom build firm houses.

Seeing our most ancient Stone arises not from combustible things, because it is free from all danger of the fire, therefore seek not for it, in such things, wherein Nature will not have it to be found or to be, as if one should tell thee it is a vegetable work, it is not, although a vegetative nature be in it.

For note, if it should be with our Stone, as it is with an herb it would easily be consumed in the fire, and nothing would remain but its Salt, and although those before me, have written many things of the vegetable Stone, yet know, my friend, that it will be difficult for you to understand it, for because our Stone, does vegetate, and multiply itself, therefore have they called it vegetable.

Know further, that brute Animals have no increase but in their like nature; therefore, need you not search after, nor presume to make the true Stone, but of its own proper seed, whereof our Stone has been made from the beginning: Also, my friend, take notice and understand, that you take not any Animal Soul for this work. For flesh and blood, as they are granted and bestowed by the Creator upon Animals, do properly belong unto Animals wherewith God has framed then, so that an Animal is made thereof: but our Stone which from the Ancients came to me as an Inheritance, proceeds and arises from two and from one thing, which contains a third concealed, this is the pure truth and rightly spoken, for male and female by the Ancients were taken for one body, not by reason of the outward appearance to the eye, but in respect of that love implanted, and from the beginning infused into them, by the operation of Nature, that they may be known to be one, and as the two do propagate and increase their Seed, so also the seed of the Matter, whereof our Stone is made, may be propagated and augmented.

If you are a true lover of our Art, you will much esteem and wisely consider this saying, lest you fall and slip with other blind Sophisters into the pit prepared by the enemy.

My Friend, that you may understand from whence this seed comes, enquire of yourself to what end you would prepare the Stone, then will it be manifest unto you, that it proceeds from no other matter, then from a certain metallic root, from whence also, the metals themselves, by the Creator, are ordained to proceed, which how it is done, Note,

That in the beginning, when the Spirit moved upon the waters, and all things were covered with darkness, then the omnipotent and eternal God, whose beginning and whose wisdom without end, was from eternity, by his unsearchable Counsel, did create the Heaven and the Earth, and all things visible and invisible in them contained, out of nothing, by what names so ever they are called, for God made all things of nothing. But how this most glorious Creation was done, I shall not now treat, let the Scriptures and Faith judge thereof.

The Creator in the Creation, gave to every Creature a peculiar seed, that there should be an increase (lest they should tend to a conclusion or detriment) whereby Men, Animals, Vegetables, and Metals might be preserved, neither is it lawful to Man, to produce a new seed at his pleasure, but is against God's Ordinance, for to him is granted propagation and increase; for the Creator has reserved to himself the power to create Seed, else were it possible for Man to act as Creator also, which must not be, but is proper to the highest power.

Then conceive thus of the seed proceeding from Metals, that the Cœlestial influence, according to Gods good pleasure and ordinance, descends from above, and mixes itself with the Astral properties, for when such conviction happens, then these two beget an Earthly substance, as a third thing, which is the beginning of our seed, its first original, whereby may be demonstrated the antiquity of its generation, from which three the elements do arise and proceed, as water, air[17] and Earth, which work further by a subterranean fire, until it bring forth a perfect thing, which *Hermes*, and all others before me, have called the three first Principles, because we could find no more from the beginning of the Magistery, and they are found to be an intrinsic Soul, an impalpable Spirit, and a corporeal and visible Essence.

Now when these three do dwell together, they do proceed by copulation, by success of time, by *Vulcans* help into a palpable substance, *viz.* into Mercury, Sulphur, and Salt, which three if by commixtion, they are brought to induration and coagulation, as Nature does many ways operate, then is there made a perfect body, as Nature would have it, and its seed is chosen and ordained by the Creator.

Whosoever you are that presumes to dive into the fountain of our work, and hopes to obtain, by your ambitious enterprise, the reward of Art, I tell you by the eternal Creator, for a truth of all truths, that if there be a Metallic Soul, a Metallic Spirit, and a Metallic form of body, that there must also be a Metallic Mercury, a

---

[17] Here the printed edition of 1670 used the word "agree," which is obviously a printer's typesetting error. From Volume 1 of The R.A.M.S Library of Alchemy, I have substituted the word "air." -pnw

Metallic Sulphur, and a Metallic Salt, which of necessity can produce no other than a perfect Metalline Body.

If you do not understand this that you ought to understand you are not adepted for Philosophy, or God conceals it from you.

Therefore, in brief thus, it will not be possible for you to attain this end with profit in a Metallic way, unless you conjoin the said three principles into one, without error. Understand further, that Animals are composed of Flesh and Blood, even as Man is, and have a living Spirit, and breath infused in them, which they enjoy as Man does; but they are without a rational Soul, wherewith Man is endued above all Animals: Therefore, when they die they are at an end, neither is there any hope of them forever. But Man, if he offers up his life by a temporal death to his Creator, his Soul survives, and after his purification, his Soul returning to his purified body shall again dwell therein, so that Body and Spirit are again united, and will clearly manifest their Cœlestial clarification, which can never be separated to all eternity, *etc*.

Therefore Man by reason of his Soul is esteemed a fixed Creature (although he dies a temporal death) yet shall he live forever; for Man's death is only a clarification, that by certain degrees ordained of God he might be freed from his grievous Sins, and transplanted into a better state, which happened not to other Animals, therefore are they not esteemed fixed Creatures, for after their death, they enjoy no Resurrection, for they want a rational Soul, for which the only and true Mediator the Son of God has shed his blood.

A Spirit may abide in some certain Body, but it does not therefore follow, that it is there to be fixed, although that Body agree with the Spirit, and the Spirit be not angry with the Body, for they both want that strong part which overcomes and confirms the Body and Spirit, and preserves and defends it from all dangers, *viz.* the most precious, noble, and fixed Soul: For where the Soul is wanting there remains no hope of Redemption; for anything without a Soul is imperfect, which is one of the highest Mysteries which ought to be known to the wise and diligent seeker of our Work: And my conscience will not suffer me to pass over this in silence, but to reveal it to those, who love the foundation of wisdom. There my beloved friend, be thou attentive to what I shall tell you, that the Spirits hid in Metals are not alike, the one being more volatile, or more fixed than the other. So also are their Souls and Bodies unequal; whatsoever Metal contains in itself all the three parts of fixity, that Metal has obtained that power to abide in the fire, and overcome all its enemies, which is only found in *Sol*: *Luna* contains in itself a fixed Mercury, therefore she flies not so soon in the fire, as do the other imperfect metals, but abides her examine in the fire, and manifests it very nobly by her victory, that greedy *Saturn* cannot prey upon her.

Amorous *Venus* clothed and possessed with an abundant tincture, for her body is almost all a mere tincture, like in color to that which is in the best metal, and by reason of its abundance of tincture appears to be red, but by reason her body is leprous, that firm permanent tincture cannot abide in an imperfect body, but is found to fly with the body; for when the body is consumed, the soul cannot stay, but is forced to be gone and fly, because its habitation is consumed and destroyed by the fire, so that it can find no place nor knows where to tarry, but in a fixed body she willingly and constantly inhabits.

Fixed Salt has given and left with warlike *Mars* a hard, constant, and gross body, whereby is manifested the generosity of his mind, from which warlike Captain can hardly anything be gotten, for his body is so hard that it can hardly be penetrated; but if his fierce valor be spiritually united with the fixity of *Luna*, and the beauty of *Venus* by a right mixture a curious harmony may be made, by which some Keys may be so advanced, that the needy if he get up the highest step of the Ladder, may get a living *Particulariter*, for the Phlegmatic quality, or moist nature of *Luna* ought to be dried up by the ardent blood of *Venus*, and its great blackness corrected by the Salt of *Mars*.

There is no necessity for you to seek your seed in the Elements, for our seed is not put so far back, but there is a nearer place, where our seed has its certain habitation and lodging, so that if you only purify the Mercury, Sulphur, and Salt, (of the Philosophers) so that of their Soul, Spirit, and Body be made an inseparable conjunction, which may never be separated the one from the other, nor can be divided, then is made the perfect bond of Love, and a habitation is sufficiently and excellently prepared f or the Crown.

Know also, that this is only a liquid Key, like unto the Cœlestial property and dry water, addicted to an Earthly substance, which are all but one thing, proceeding and growing from three, two, and one, if you can apprehend this, then have you obtained the mastery, then conjoin the Bride with the Bridegroom, that they may feed and nourish each other with their own flesh and blood, and increase infinitely from their own seed.

Although I could willingly out of love reveal more unto you, yet the Creator has prohibited me; wherefore it becomes me not to speak more clearly of these things, lest the gifts of the Most High be abused, and that I should be the cause of committing of many sins, so that I should pull down Divine Vengeance upon me, and with others be cast into eternal punishments.

My friend, if these things be not clear enough unto you, then will I lead you to my practick part, which I accomplished, the Stone of the Ancients by the assistance of the Almighty, consider it well, and with diligent and frequent reiteration thoroughly read my XII Keys, and so proceed, as I shall here teach and instruct you, fundamentally by way of Parable.

Take a piece of the best fine Gold, and separate the same in parts, by such means as Nature has granted unto the Lovers of Art, even as an Anatomist divides the dead body of man, and thereby searches into the inward parts of the human body, and make your Gold to be reduced to what it was at the first, then will you find the Seed, the beginning, middle, and end, whereof our Gold and its Wife were made, *viz.* out of a penetrating subtle Spirit, and of a pure chaste and immaculate Soul, and of an Astral Salt and Balsam, which after their conjunction are nothing else but a Mercurial Liquor, which same water was brought to School to its own God *Mercury*, who examined that water, and having found it to be legitimate and without deceit, he joined in friendship with it, and joined with it in Matrimony, and so of both them was made an incombustible Oil, then Mercury grew so proud that he scarce knew himself; he cast off his Eagles wings, and himself swallowed up the slippery tail of the Dragon, and offered battle to *Mars*.

Then *Mars* gathered his Champions together, and gave command that Mercury should be imprisoned, to whom *Vulcan* was appointed Gaolor, until he should be freed by some of the feminine kind.

After these things were rumored abroad, the other Planets met together, and held a counsel, they considered what was fit to be done that they might wisely proceed. Then *Saturn* the first in order with a fierce speech began to speak after this manner.

I, *Saturn*, the highest Planet in the firmament, protest before you all my Lords, that I am the most unprofitable and contemptable of you all, of an infirm and corruptible body, of a black color, obnoxious to the injuries of many afflictions in this miserable world, yet am the examiner of you all. For I have no abiding place, and I take with me whatsoever is like unto me: the cause of this my misery is to be imputed to none but to inconstant *Mercury*, who by his carelessness and negligence has brought this evil upon me: Therefore, my Lords, I pray you, revenge my quarrel on him, and seeing that he is already in Prison, kill him, and let him putrefy there, until not one drop of his blood be any more found.

Saturn having ended his Speech, brown *Jupiter* came on, and began his speech, with his bended knees, and with the reverential honor of his Scepter, commending the requests of his fellow *Saturn*, commanded all such to be punished that should not put those things in execution and so he made an end.

Then came *Mars* with his naked Sword variously colored, like a fiery glass, shining with divers and strange rays, he brought this sword to *Vulcan* the Gaolor, to put therewith in execution all those things commanded by the Lords, which when he had killed *Mercury*, he burnt his bones in the fire, wherein *Vulcan* the Gaolor was very obedient.

In the meantime, whilst the Executioner was performing his office, comes in a beautiful and white shining Woman, in a long Robe of a silver color, woven with several water-colors which when she was received, she appeared to be *Luna, Sols* wife; she fell on her face, and with many tears, and on her knees besought them that her husband *Sol* might be set at liberty out of the Prison, into which *Mercury* by force and deceit had cast him in, where to this day he has been detained by the Command of the other Planets; but *Vulcan* denied her, for he was so commanded to do, and persisted. in his purpose in executing the Sentence. Then came dame *Venus* in a garment of pure red, interwoven with green, of a most beautiful countenance, a most graceful and pleasant speech, and most amiable gesture, bearing most fragrant flowers in her hand, which by the variety of the colors did wonderfully refresh and delight the eyes of those that looked on her; she made intercession in the *Chaldean* Language unto *Vulcan* as Judge for liberty, and put him in remembrance, that Redemption must come from a woman kind; but his ears were stopped.

In the meantime, while these two thus conferred together, the heaven opened itself, and thence came a great Animal with many thousands of young ones, driving away and expelling the Executioner. He opened his Jaws wide, devoured the precious Lady *Venus* the Interceder, crying with a loud voice, my descent is of Women, and Women have plentifully spread abroad my seed, and have filled the Earth with it; her Soul is kind to me, therefore will I feed and nourish myself with her blood: When this Animal had thus loudly spoken, he withdrew himself into a certain Conclave, and shut the door after him, and all his young ones followed him in order, where they wanted much more food than before, and they drank of the former incombustible Oil, and they did easily digest their meat and their drink, and they had many more young ones than before, and this happened often, until they had replenished the whole world.

When all these things had so happened, many skillful men of every Country, learned in all kind of Studies, met together, who endeavored to find out the interpretation of all those things and speeches, that they might for the most part better understand those things, but none of them could attain unto it, for they were not all of one mind, until at last came forth a certain Old Man, his Beard and Hair as white as Snow, with a Purple garment from the Head to the Foot, he had a Crown on his Head, whereon there shined a most precious Carbuncle, he was girt about with the Girdle of Life, he went on his bare Feet, he spoke from a singular Spirit that was hid in him, his speech penetrated through the innermost parts of the body, so that the Soul heartily received it. This man ascended the Chair, and exhorted the Assembly there met to be silent, and to harken diligently to what he should tell them, for he was sent from above to interpret unto them the aforesaid writings, and to reveal it by Philosophic expressions.

When they were all quiet, he began after this manner.

Awake, O Man, and contemplate on the Light; lest the Darkness seduce you: The Gods of Fortune, and the Gods of the greater Nations, have revealed unto me in a deep sleep. O how happy is that Man, that acknowledges the Gods, how great and wonderful things they work, and happy is he whose eyes are opened, that he may see the light, which before was hidden.

Two Stars have the Gods granted unto Man, to lead them to great wisdom, which steadfastly behold, O Man, and follow their Splendor, for wisdom is found in them.

The Phoenix of the South has snatched away the heart out of the breast of the huge beast of the East, made wings for the beast of the East, as has the bird of the South, that they may be equal; for the beast of the East must be bereaved of his Lions skin, and his wings must vanish, and then must they both enter the Salt Ocean, and return again, with beauty. Sink your disquieted Spirits into a deep fountain, that never wants water, that they may be like their Mother, that lies hid therein, and from three came into the world.

Hungary first begot me, the Heaven and the Stars preserve me, and I am married to the Earth; and although I am forced to die, and to be buried, yet *Vulcan* revives me the second time, therefore *Hungary* is my Native Country, and my Mother contains the whole world.

When these things were heard by the Assembly then present, he further spoke thus.

Make that which is above to be beneath, and that which is visible to be invisible, and that which is palpable to be impalpable: And again, make that which is above be made of that which is beneath, and the visible of the invisible, and the palpable of the impalpable thing; thus is the whole Art absolutely perfect without any defect or diminution, wherein dwells Death and Life, Death and Resurrection, it is a round Sphere, wherein the *Goddess of Fortune* drives her Chariot, and communicates the gift of Wisdom to the Men of God. Its proper name is according to our temporal understanding, *All* in *All*; The Highest he is Judge over things eternal.

Whosoever desires to know what the *All* in *All* is, let him make very great Wings for the Earth, and force her so much, that she lift herself up, and raise herself on high, flying through the Air into the Supreme Region of the highest Heaven. Then burn her wings with a very strong Fire, that the Earth may fall headlong into the Red Sea, and be drowned therein, and with Fire and Air dry up the Water, that thereof Earth may be made again; then I say have you the *All* in *All*.

But if you cannot apprehend this, inquire into thyself, and seek about in all things that are to be found throughout the world: Then will you find the *All* in *All*, which is the Attractive Power of all Metallic and Mineral things, proceeding from *Salt* and *Sulphur*, and twice begotten of *Mercury*: More (I tell you) is not meet for me to speak of that, which is the *All* in *All*, because *All* is comprehended in *All*.

This Speech being made, he said further, O my Friends, thus by the hearing of my voice have you learned wisdom, from what and by what means you ought to prepare the Great Stone of the Ancient Philosophers, which heals all leprous and imperfect Metals, and opens unto them a new Birth, and preserves men in health, and prolongs their lives, and has hitherto preserved me by its Cœlestial power and operation, that I am very willing to die, being weary of this life.

Praised be God forever for his grace and wisdom, which of his mercy he has a long time bestowed on me, *Amen*.

And so he vanished away before their eyes.

This speech being ended, everyone returned unto his own home from whence they came, meditating night and day on those things, and laboring every one according as the expertures of their Genius enabled them, *&c*.

Now follow the

XII. Keys of Basilius Valentinus

**Wherewith the Doors are opened to the most Ancient
Stone of our Ancestors, and the most secret
Fountain of all Health is discovered.**

The I. Key.

Know, my Friend, that impure and defiled things are not fit for our work; for their Leprosy, can be no help in our operations; that which is good is hindered by that which is impure.

All wares sold from the Mines are worth one's money, but when they are sophisticated, they are unfit for use, for they are counterfeited, and are not of the same operation as they were before.

As Physicians cleanse and purify the inward parts of the body, by means of their medicines, expelling all impurities from thence. So also ought our bodies to be purged and purified from all their impurities, that perfection may be wrought in our Birth: Our Masters require a pure and undefiled body, which is not adulterated with any spot or strange mixture: For the Addition of another thing is a Leprosy to our Metals.

The Kings Diadem is made of pure Gold, and a chaste Bride must be married unto him.

Wherefore if you will work upon our bodies, take the most ravenous grey Wolf, which by reason of his Name is subject to valorous *Mars*, but by the Genesis of his

Nativity he is the Son of old *Saturn*, found in the Mountains & in Valleys of the World: He is very hungry, cast unto him the Kings body, that he may be nourished by it; and when he has devoured the King, make a great fire, into which cast the Wolf, that he be quite burned, then will the King be at liberty again: When you have done this thrice, then has the Lion overcome the Wolf, neither can he find any more on him to feed upon; and so is our body prepared for the beginning of our work.

Know also that this is the right and true way to purge our bodies, for the Lion purifies himself by the blood of the Wolf, and the tincture of his blood wonderfully rejoices in the tincture of the Lion, for both their bloods are nearer of Kin one to the other; when the Lion is satisfied his Spirit is made stronger than it was before, and his Eyes shine with great splendor like the Sun, and his inward Essence is of great efficacy, and is profitable for anything you apply it unto: And when it is so prepared, the sons of men return it thanks, who are troubled with grievous diseases, falling sickness, and other distempers: The Ten Lepers follow him, and desire to drink of the blood of his soul, and all such that are afflicted with diseases, exceedingly rejoice in his spirit.

For whosoever drinks of this golden fountain, soon feels a renewing of his nature, the taking away of evil, the comforting of the blood. The strengthening of the heart, and the perfect healing of all members throughout the body, either exterior or interior: It opens the pores, expelling the evil, that good may come in its place.

But, my Friend, you must take very diligent care, that the Fountain of Life be pure and clear, that no strange waters be mixed with our Fountain, lest it prove a miscreant, and of a wholesome Fish a Serpent be produced: If also by a medium a corrosive should be joined, by which our body might be dissolved, see that all the corrosive be washed away; for no corrosives are to be used against inward diseases, sharp things penetrate and destroy, and beget more disaster, our Fountain must be without any poison, although poison expels poison.

When a tree brings forth unwholesome and ungrateful fruit, it is cut off at the stem, and some other kind of fruit is grafted in, then the graft unites itself with the stem, so that of the stem, root, and graft, a good Tree is produced, which according to the workman's desire brings forth wholesome and pleasing fruit.

The King walks through six places in the Cœlestial Firmament, but in the Seventh he keeps his seat; for the Kings Palace is adorned with golden Tapestry: If now you understand what I say, then have you opened the first Lock with this Key, and removed the bolt that hindered, but if you cannot find any light herein, then will not your glass Spectacles profit you anything, nor your natural Eyes help you find out that at last which you would at the beginning. I shall say no more of this Key, as *Lucius Papirius* taught me.

The II. Key.

In the Court of the great Potentates various kinds of drink are found, yet scarce any of them are alike in smell, color, and taste, for their preparation is different; yet all they all drink, because they are all made and necessary for their particular uses in the family.

When the Sun sends forth his beams, irradiating them through the Clouds, it is commonly said, that the Sun attracts Water, and that it will rain; and that if it often happens, the year proves fruitful.

For the building of a Princely Palace, various and divers Workmen and Mechanics must be set on work, before it be called a beautiful and perfect Palace. Where stones are required, wood must not be used.

Through the daily ebbing and flowing of the raging Sea, which are caused by a certain Sympathy from the Cœlestial influences, Countries are enriched with many and great riches, for at every return it brings with it some good to the Inhabitants.

A Virgin that is to be married, is first richly adorned with variety of precious garments, that she may please her Bridegroom, and beget him by looking on her a more vehement affection, but when the Spouse is to take a carnal cognizance of her

Husband, all these Garments are laid aside, neither does she keep anything on her, but what the Creator granted her at the beginning.

Even so our Bridegroom *Apollo*, with his Bride *Diana*, is to be married, but first diverse Garments are to be made for them, their heads and bodies must be well washed with water: which waters you must learn by the divers ways of distilling, for they are much unlike, some are strong, some are weak, according as there is use for them, as I said of the several sorts of drink; and know that when the humidity of the Earth ascends, and is elevated to the Clouds, it is there coagulated, and by reason of its ponderosity falls again, whereby the abstracted humidity is again restored unto the Earth, which refreshes, feeds, and nourishes the Earth, that leaves and grass do thence spring forth, therefore some preparations of your waters ought to be often distilled: That that which is drawn from the Earth, may be often returned unto it, and often abstracted; as the Sea *Euripus* does often leave the Earth, and covers it again, always keeping its bounds or period.

When thus the Kings Palace is prepared and adorned by several workmen, and the glassy Sea is finished, and the Palace furnished with goods, then may the King safely enter, and keep there his Residence.

But my friend know, that the naked Bridegroom must be espoused to his naked Bride; therefore all those preparations for the adorning their Garments, and beautifying their faces, must be taken away, that they may lie down as naked as they were born, that their seed be not destroyed by any strange mixture.

For a conclusion of this discourse, I tell you truly, that the most precious water, wherewith the Bridegrooms Bable[18] must be made, must be wisely and with great care prepared of two Fencers (understand of two contrary matters) that one adversary may drive out the other, and they must be prepared for the fight, and the Prize must be won: For what advantage is it for the Eagle, to build her Nest in the Rocks, where her Chickens will die on the tops of the Mountains, by reason of the coldness of the Snow?

But if you add to the *Eagle* the old *Dragon*, which has a long time had his habitation among Stones, and creeps out of the Caves, and put them both in the Infernal Pit, then will *Pluto* breathe upon them, and will enforce a fiery volatile Spirit out of the cold *Dragon*, which by its great heat burns the *Eagles* feathers, and makes a sweating Bath, that the Snow on the highest Mountains melts, and turns into water.

Whereby the Mineral Bath is well prepared, which brings riches and health to the King.

---

[18] Most likely Bath, as used in The R.A.M.S. Library of Alchemy Volume 1 version. -pnw

The III Key.

By water fire may be wholly extinguished, if much water is cast into little fire, then the fire gives way to the water, and yields up the victory unto it: So must our fiery Sulphur be conquered, and overcome by water prepared according to Art.

If after the separation of the water, the fiery life of our Sulphureous Vapor can but again triumph and obtain the victory; but no conquest can be herein obtained, unless the King add force and power to his Water, and has given it the Key of his own proper color, that he may be thereby destroyed and made invisible; yet at this time his visible form ought to return, yet with a diminution of his simple Essence, and melioration of his Condition.

The Limner can paint yellow upon white, and red upon yellow, and then a purple color; and although all the colors appear, yet the last highly excels in its degree: The like ought to be observed in our *Magistery*, which being done, then have you before your eyes the light of all wisdom, that shines in the darkness, but burns not.

For our Sulphur burns not, yet it shines far and near, neither does it tinge anything, unless it be prepared and tinged with its own tincture, whereby it may afterwards tinge weak and imperfect Metals: For it is not in the power of this Sulphur to tinge, unless the tincture be given it in the fixation: For the weaker cannot overcome, but the stronger may obtain the victory over the weaker, and the weak must yield to the strong. Therefore, observe for this discourse the following conclusion.

That which is weak cannot succor the weak, nor administer any help in the operation, and one combustible thing cannot defend another combustible thing, lest it also be burned; therefore if a defender must be, that must assist the combustible and defend it: Then that defender must have a greater power than he that needs his defense protection, and principally in its substance ought to be incombustible.

So, he that would prepare our incombustible Sulphur of the Philosophers, let him first consider with himself, that he seeks our Sulphur in that, wherein it is incombustible, which cannot be, unless the Salt Sea have swallowed up the body, and cast it up again: Then exalt it in its degree, that it far exceeds in brightness all the other Stars in the Heaven.

And in its own Essence is so full of blood, as is the Pelican, when she wounded her own breast, and without prejudice to her body, nourishes and feeds many young ones with her own blood.

This is the Rose of our Masters, of a purple color, and the Red Blood of the Dragon, whereof so many have written; it is that purple Mantle, richly leaved, in our Art, wherewith the Queen of health is covered, and wherewith all Metals wanting heat may be revived.

Keep safely this honorable Mantle, together with the Astral Salt, which follows this Cœlestial Sulphur, lest some evil befall it, and give unto it of the volatility of the Bird, as much as will suffice; then will the Cock devour the Fox, and will afterwards be drowned in the water, and being revived by the fire, will be again devoured by the Fox, that like may be restored to its like.

The IV. Key.

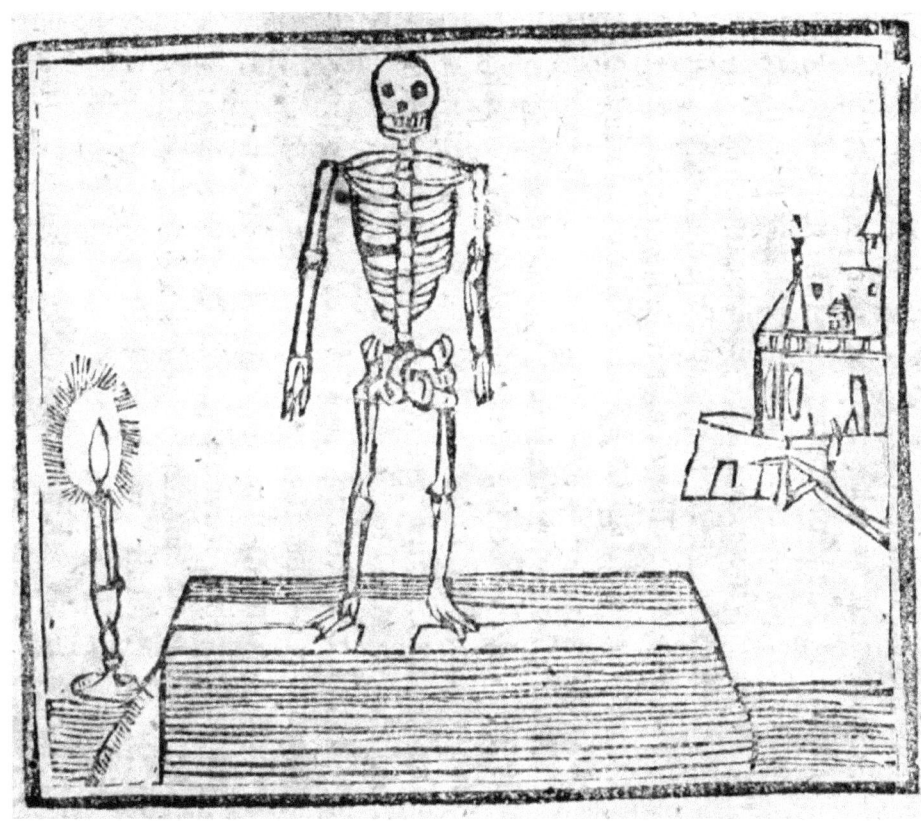

All Flesh that came from the Earth, must be corrupted and return to Earth again, as it was Earth at the first, then that Earthly Salt begets a new generation, by a Cœlestial revivification, for if it were not first Earth, there could be no revivification in our work; for in the Earth is the Balsom of Nature, and is their Salt who sought after the knowledge of all things.

At the Day of Judgement, the World shall be judged by Fire, that which was made by the Creator of nothing, must by Fire be burnt to Ashes, out of which Ashes the Phoenix produces her young: For in those Ashes lie the true and genuine Tartar which must be dissolved, and when that is dissolved, the strongest Lock of the Kings Palace may be opened.

After that burning, a new Heaven, and a new Earth shall be formed, arid the new Man shall more gloriously shine forth, then ever he lived in the old World, for he shall be purified.

When Ashes and Sand are well maturated and concocted in the fire, then the Artist turns it into Glass, which afterwards will endure in the fire, and in color like a transparent Stone, and is not any more like Ashes; and this to the ignorant is a great

Mystery, but not so in any wise to the experienced Artist, because they understand the reason thereof, by their understanding, and daily experience.

Workmen prepare Lime of Stones by burning them, that it may be fit for their use; for before its preparation in the fire, it is a Stone, and cannot be used in work as Lime: The Stone is maturated in the fire, and receives from the fire a very high degree of heat, and is made so strong, that there is scarce anything comparable to the fiery Spirit of *Calx vive*, if it be brought to its perfection.

Everything being burnt to Ashes by Art will yield a Salt, if in the Anatomizing thereof you are able to keep apart its Sulphur and Mercury, and again restore them to their Salt, according to the pure method of Art; then may you again by means of Fire, make thereof again, what it was before its destruction or anatomy; which the wise Men of the World call foolishness, and esteem those things as trifles, and say, this is a new Creation, which God grants not to sinful man, but they do not understand that this was created before, and that the Artist does only show its increase and Magistery by the Seed of Nature.

If the Artist want Ashes, he cannot make Salt for our Art, for without Salt our work cannot be made into a body, for Salt only coagulates all things.

For as Salt does sustain all things, and preserves them from putrefaction; even so the Salt of our Masters preserves Metals, lest they be reduced to nothing, and be corrupted, which can in no wise happen, unless their Balsom perish, and the incorporated Saline Spirit cease to be; then would their body be altogether dead, and nothing could be thereof made to any advantage, because the Spirits of the Metal are decayed, and at their departure left a naked and void habitation, into which no life can be again restored.

You that are Students in this Art, know further, that Salt out of Ashes is of very great use, much virtue is contained in them, yet is that Salt unprofitable, unless its inside be turned outwards, and its outside inwards, for it is the Spirit only that gives power and life, (for the naked body avoids nothing). If you know how to obtain that, then have you the Salt of the Philosophers, and the true incombustible oil, whereof they have written many things before me:

> Although that many wise,
> Have sought for me with care;
> Yet few consider what,
> My hidden treasures are.

The V. Key.

The vivifying power of the Earth, produces all things that proceed therefrom: And he that says that the Earth is without life, is in an error.

For that which is dead cannot add to that which has life, and the off-spring of the dead cease, because the Spirit of Life is wanting; therefore, the Spirit is the Life and Soul of the Earth, that dwells in it, and operates on Earthly things, from the Cœlestial and Sidereal. For all Herbs, Trees, and Roots, and all Metals and Minerals, receive their powers, increase, and nourishment from the Spirit of the Earth: For the Spirit is the Life, which is nourished by the Stars, and administers of its nourishment to all Vegetables. And as the Mother preserves the *Fetus* in the Womb, and feeds it there; so also does the Earth nourish in its bosom the Minerals, by its Spirit received from above.

Wherefore the Earth affords not those virtues of itself, but the living Spirit which is in it; and if the Earth should be without that Spirit, it was dead, and could not yield any more nourishment, because the Spirit would want that Sulphur or

Fatness which preserves the vivifying power, and produces all growing things by its Nutriment.

Two contrary Spirits may dwell together, but not easily agree. For when Gunpowder is kindled, these two Spirits whereof it is made, fly the one from the other with great noise and violence, and vanish into fume, that none knows whither they are gone, or what they were, unless they knew by experience what Spirits they were, and in what subject they had had their being.

Whence you may know, thou searcher into Art, that Life is only a mere Spirit, so that everything that the ignorant World repute for Dead, may be again reduced into an imperceptible, visible, and Spiritual life, and may be preserved therein, if Life only can operate with Life, which Spirits feed and nourish themselves by a Cœlestial substance, and are procreated from a Cœlestial, Elementary, and Terrestrial substance, which is called the *Materia informis*.

And as Iron has its Magnet, which by its wonderful and invisible love attracts it; so also has our Gold a Magnet, which Magnet is the *Prima Materia* of the Great Stone. If you understand these my expressions, you are blessed with riches above all the world.

I will reveal one thing more unto you in this Chapter. When a Man looks in a Glass, there is the reflection of his Image, which if you go to touch with your hands, you find nothing tangible but the Glass wherein the person looked: So also from this matter must be drawn a visible Spirit, which nevertheless is impalpable. That very same Spirit, I say, is the Radix of the Life of our Bodies, and the Mercury of the Philosophers, from whence our liquid Water is prepared in our Art, which you must make again Material with its own Composition, and by some certain means reduce it from the lowest to the highest degree into a most perfect Medicine. For our beginning is a secret and palpable body, the middle, is a fugitive Spirit, and a golden water without any corrosive, by which our Masters prolonged their lives; and the end is a most fixed Medicine for human and Metalline bodies, which to know is rather granted to Angels than Men, although some men are adopted to the same knowledge, who by their earnest prayers obtained the same of God, and are thankful to him therefore, and helpful to the needy.

For a Conclusion to these things, I tell you for a truth, that one work proceeds from another, for our matter must be very well and highly purified in the beginning of our work, then dissolved and destroyed, and thoroughly broken and reduced into dust and ashes: When this is all done, then make thereof a volatile Spirit, white as Snow, and another volatile Spirit red as Blood, which two Spirits contain in them a third, and yet are but one Spirit. These are the three Spirits, that preserve and prolong Life, join them together, give them their natural meat and drink, as much as they need, and keep them in a warm bed, until the perfect time of their Nativity; then shall you see and understand what the Creator and Nature has discovered unto you; and know that my lips never yet so plainly revealed anything. For God has placed more efficacy and wonderfulness in Nature, than many thousands of Men can believe; but I am sealed upon also that others after me may write of those wonderful natural things which are granted by the Creator, but by fools are counted for supernatural; for that which is natural has its first original from that which is Supernatural, and yet are they found together to be only Natural.

The VI Key.

Man without a Woman is esteemed but as half a body, and a Woman without a Man likewise obtains the name of half a body, for either of them by themselves can produce no fruit; but when they live together in a Conjugal State, the body is perfect, and by their Seed an increase succeeds.

When too much seed is cast on the ground, that the land is overburdened, mature fruit cannot be expected; and if there be too little seed then the fruit comes up thin, and instead thereof grow tares, whence no profit can be expected.

If any will not burden his Conscience with sin in selling of Wares, let him give his Neighbor just measure, and let him use just weights and measures, then he avoids curses, and gains the blessings of the poor.

In great waters, it is easy to be drowned, and shallow waters are easily exhausted by the heat of the Sun, that they are of no use.

Therefore to obtain your desired end, a certain measure must be observed in the commixtion of the Philosophic Liquid Substance, that the greater part do not over-power and over-press the lesser, whereby the Effect will be hindered, and lest the lesser be too weak for the greater, but let there be made an equal dominion: For great rains are unprofitable to fruits, and overmuch drought hinders true Maturity, wherefore if *Neptune* has rightly prepared his Water-bath, then take a just quantity of the *Aqua Permaneus*, and have a great care that you take not too much, nor too little.

A double fiery man must be fed with a white Swan which will kill each other, and will again revive. And the Air of the four parts of the World must possess three parts of the included fiery Man, that the song of the Swan may be heard, when she harmoniously sings her farewell, Then the roasted Swan will be food for the King, and the fiery King will exceedingly love the pleasant voice of the Queen, and out of his great love embrace her, and satiate himself with her, until both vanish and become one body.

It is commonly said, that two can overcome and conquer one, especially if there be space enough to exercise their fury; wherefore know from a true ground, that a double wind must come called *Vulturnus*, and then a single wind called *Notitus*, these will fiercely blow from the East and South, but when they cease, so that of the Air is made Water, then be confident that a Corporeal thing may be made of a Spiritual, and that the number will bear rule through the four parts of the year, in the fourth Heaven, after the Seven Planets have exercised their dominion, and will finish their course in the lowermost habitation of the Palace, and are ready for the highest Examin, so those two that were sent have overcome and consumed the third.

The Knowledge of our Magistery is herein very necessary for division, and conjunction must be rightly made, if Art is to produce riches, and the Scales must not be falsified by unequal weights. This is the Rock we proposed, that you be sure to finish this Work by an artificial Heaven, by Air, and by Earth, with true Water and perceptible Fire, in giving of a lawful weight without any defect, as I have rightly informed you.

The VII. Key.

Natural heat preserves the Life of Man, for if that be gone, Life ceases.

Natural fire, if it be moderately used, defends against cold, but too much is destructive. It is not of necessity that the Sun do corporally touch the Earth, it is sufficient that the Sun does manifest its power at a distance by its Rays, which by reflection against the Earth, are much strengthened, for by that means it has power enough to perform its office, and to maturate all things by concoction; for by the distance of the Air the Sun-beams are temperated, so that by means of the Air the Fire operates, as by means of the Fire the Air operates.

The Earth without the Water can produce nothing, and again, the Water without the Earth excites nothing; as the Earth and the Water need each other's assistance in the production of fruits, so in no wise can the Fire without the Air, nor the Air without the Fire, for the Fire without the Air has no Life, and the Air without the Fire cannot manifest its heat and dryness.

The Vine has more need of the heat and beams of the Sun at the latter time of its maturation, than it had in the beginning of the spring: And if the Sun strongly operates in the Autumn, the Vine yields a better and stronger Juice, than if the heat of the Sun-beams be weak or deficient.

In the Winter the common people count all things dead, because the cold binds the Earth that nothing can grow, but as soon as the Spring appears, that the cold lessens by the ascent of the Sun, all things revive, Trees and Herbs grow, and Insects which hid themselves from the cold Winter creep forth out of their holes and caves in the Earth; all Vegetables yield a new savor, and their Excellency is discovered by their fair, amiable, and various colors of their blossoms; and then the Summer continues the operation, and brings forth Fruits from these several kinds of Flowers: For which thanks be given to the Creator, who by his Ordinance has set bounds unto these things by Nature.

So year follows after year, until the World be again destroyed by its Maker, and they that Inhabit therein be exalted by the glory of God, then shall all Earthly Nature cease to work, and the Eternal Cœlestial one shall be in its stead.

When the Sun declines from us in the Winter, it cannot dissolve the Snowy Mountains, but when it approaches nearer in the Summer, the Air is hotter, and more powerful to dissolve the Snow, that it turns it into water, and destroys it: For the weak must yield to the stronger, and the stronger overrules the weak.

Thus also in our Magistery the government of the Fire must be observed, that the moist Liquor be not too suddenly dried up, and the Philosophic Earth too suddenly melted and dissolved, else out of wholesome Fishes in your water you will generate Scorpions. But if you desire to be a true Master of your Work; Then take your Spiritual Water, wherein the Spirit moved at the beginning, and shut the door of defense upon it; for from that time shall the Heavenly City be besieged by Earthly Enemies, and your Heaven must be strongly defended with three fences and walls, that there be no entrance but one, and let that be very well guarded. When all these things are done, kindle your Philosophic Lamp, and seek what you have lost, give so much light as may suffice: For know, that Insects and Worms die in the cold and moist Earth, for it is their Nature, but man's habitation is ordained to be upon the Earth in a temperate and even condition: but the Angelical Spirits that have not an Earthly, but an Angelical body, and are not obnoxious to the pollutions of sinful Flesh, as man is, they are placed in a higher degree, that they can bear without any prejudice both heat and cold, both in the higher and lower Region: And when nan shall be purified, he shall be like those Cœlestial Spirits; for God rules both Heaven and Earth, and works all things in all.

If we rightly behold our own Souls, then shall we be made Sons and Heirs of God, to effect that which seems now impossible to us: But this cannot be done unless the Waters be dried up, and Heaven and Earth with all Men be judged by Fire.

The VIII. Key.

All flesh be it Man's or Beast's yields no increase or propagation, unless it be first putrefied, also the seed when it is sown, and all that is under or belonging to Vegetables cannot increase but by putrefaction.

Many insects and worms receive life, so that by mere putrefaction they attain a vivifying power and motion; which ought to be deservedly esteemed, as a wonder above all wonders: This Nature has granted, for the same vivifying increase and inspiration of life is very much found in the Earth, and by the same reason is excited in its Spiritual seed by the other Elements.

This is demonstrable by examples; the Country-wife knows it very well, for they cannot produce a Hen for their use, but by the putrefaction of the Egg, out of which the Chicken is generated.

If bread fall into honey, Ants are bred there, which is a singular mystery in Nature above others: The Country-man also understands, that worms proceed out of putrid flesh of Men, Horses, and other Beasts, and also Spiders, worms, etc. in Nuts, Apples, Pears, &c. None are able to enumerate the various kinds and species of Worms, which proceed from putrefaction.

The same is also observed in Vegetables, that divers kinds of Herbs, as Nettles arid many others grow in those places where such never grew, now their seed fell, only came by putrefaction, the cause is, that the earth in those places is disposed, and as it were impregnated for these productions, which has been infused from above by the Sidereal property, then the Seed has spiritually been formed into them; which Seed putrefies itself in the Earth, and by the operation and co-assistance of the Elements does generate a corporeal matter, according to its natural species; so that the Stars with the Elements can excite a new Seed which was not before; and afterwards by succeeding putrefaction may be increased; But it is not granted to Man to excite a new Seed; for the Operation of the Elements, and the Essence of the Stars are not in his power to form.

Thus divers sorts of Vegetables grow only by putrefaction, but the Country-man looks upon it as usual, and considers not, neither can he imagine or understand any reason for it, for by them its esteemed only as customary: But you, whom it becomes to know more than the vulgar, may learn the causes and fundamentals by observing these my large demonstrations and expressions, *viz.*, from whence this living power of resuscitation and generation should proceed, not esteeming it as customary, but of a diligent searcher into Natures Mysteries; because in truth all Life proceeds from and is caused by putrefaction.

Every Element has in itself its corruption and its vicissitude of generation: Let the desires of Art be sure of this, and know it from a right foundation, that in every Element the other three are hid: For the Air contains the Fire, Water, and Earth in itself, which seems to be incredible, yet it is true; so also the Fire contains the Air, Water, and Earth; and Earth contains Water, Air, and Fire, else they could not generate.

And the Water has part of the Earth, Air, and Fire, otherwise no generation could follow, yet notwithstanding every Element is distinct, although they are all mixed, which is evidently found by distillation in the separation of the Elements.

But I will more clearly demonstrate this to you, lest you being ignorant, judge that what I have said are mere words and not truths, I tell you, who earnestly intend the separation of Nature, and to understand the division of the Elements, that in the distilling of the Earth, first the Air comes very easily, then after some certain time the Element of Water, the Fire was included in the Air, because both are a Spiritual Essence, and do both wonderfully love each other. The Earth remains in the bottom, wherein is the most precious Salt.

In the distillation of the Water, the Air and Fire first come over, then the Water and the body of the Earth remains in the bottom. The Element of Fire, if it be extracted by fire into a visible substance, the Water and the Earth may be taken apart, so then the Air remains in the other three Elements, for none of them can want Air: The Earth is nothing, neither can it produce anything without Air; the Fire burns not, nor has any life without Air; the Water cannot bring forth any fruit without Air, neither can the Air consume anything, nor dry up any moisture, but by natural heat, because fervor and heat is found in the Air, therefore the Element of Fire must needs be in the Air; for whatsoever is hot and dry is of the substance of Fire; wherefore one Element cannot want the other, but the commixtion of the four Elements is always found in the generation of all things: And he that denies this does not understand the Mysteries of Nature, neither has he searched into their properties.

For you ought to know, if anything proceed by putrefaction it must of necessity be after this manner: The Earth by its secret and hidden moisture is reduced into corruption, or a certain destruction, which is the beginning of putrefaction; for without moisture, as the Element of Water, there can be no true putrefaction: For if any generation do proceed from putrefaction, it must needs be kindled and produced by the property of heat or Element of Fire; for without natural heat no production can be made, and if that production do assume a living breath and motion, that cannot be without Air; for if the Air did not co-operate therewith, and lend its aid, then the first composure and substance from whence the generation proceeded, would of itself be suffocated and die for want of Air; wherefore it is clearly seen, and fundamentally demonstrated, that no perfect creature can be generated without the ministration of the four Elements, and that always one Element shows its operation and life in another, which is manifested by putrefaction.

For without that (*i.e. Air*) nothing could be brought to light, from this time and forever, and that all the four Elements are necessary for a perfect generation, and resuscitation. Know that when *Adam* the first Man was formed by the great Creator out of a lump of Earth, there did not as yet appear any perceptible motion of life, until God breathed a Spirit into him, then was this lump of Earth endowed with power. In the Earth was the Salt, *i.e.* the body, the inspired Air; with the Mercury, the Spirit. The Air by this inspiration did give a genuine and temperate heat, which was Sulphur, *i.e.* Fire, then it moved itself, and *Adam* manifested by this motion, that a living Soul was inspired into him, for Fire cannot be without Air, and so no Air without Fire: The Water was incorporated with the Earth, for of necessity they must be together in an equal commixture if you expect life to follow.

So *Adam* was first brought forth, generated and compounded of Earth, Water, Air, and Fire; of Soul, Spirit, and Body; and of Mercury, Sulphur, and Salt.

After the same manner *Eve*, the first woman and Mother of us all, partook of the same composition, being taken from *Adam*; so *Eve* was produced and built from *Adam*: which note well.

And that I may again return to putrefaction, let the seeker of our Magistery, and inquirer into Philosophy know, that for the same reason no Metalline seed can operate, or augment itself, unless this Metalline seed, by itself only, without any strange addition or mixture, be brought to a perfect putrefaction: As no Seeds of Vegetables or Animals can produce any increase without putrefaction, as is already declared; so also understand of Metals, which putrefaction must attain its perfect operation by the benefit of the Elements; not that the Elements are the Seed, as before is sufficiently manifested, but that Metallic seed, which is begotten by a Cœlestial, Sidereal, and Elementary Essence, and is brought into a Corporeity, must be further reduced, by the Elements into such a putrefaction and corruption.

Note this also, that Wine contains a volatile Spirit, for in its distillation the Spirit first comes over, and then the phlegm: But if it be first by continual heat turned into Vinegar, its Spirit is not so volatile as before; for in the distillation of the Vinegar, its Water or phlegm comes first, and the Spirit last; and although it be the same matter that was before in the vessel, yet has it by far another property, being no more wine, but by the putrefaction of continual heat is transmuted and made Vinegar: And everything that is extracted and circulated with Wine, or its Spirit, has a far different property and operation, then that which is extracted with Vinegar.

For if the Vitrium of Antimony be extracted with wine, or Spirit of Wine, it provokes many Stools and Vomits; because its poison and venom is not yet destroyed or extinct: But if the Vitrium of Antimony be extracted with good distilled Vinegar, it gives a fair extraction of a high color: Then extract the Vinegar *per* M.B., and the yellow remaining powder being well edulcorated by often washings with common water; that all acetosity be done away, then is it a sweet powder which does not excite any more Stools, but is a very excellent Medicine for use, which exceeds even to admiration, and may be deservedly called the Wonder of Medicine.

This wonderful powder in a moist place resolves into a Liquor, which is of excellent use in Chirurgery, curing without any pain.

Whereof enough.

And this is principally to be noted for a conclusion of this discourse, that there is a Cœlestial Creature generated, whose life is preserved by the Stars, and fed by the four Elements, which ought to be killed, and then putrefied, which done, the Stars by means of the Elements will again infuse life into those putrid bodies, that it may again be made that heavenly substance, which had its habitation in the highest Region of the Firmament, if that be done, you shall perceive that the Terrestrial body is reduced into a Cœlestial Substance.

The IX. Key.

Saturn the highest of the Cœlestial Planets, has the meanest authority in our Magistery, yet is he the chiefest Key in the whole Art; but placed in the lowest degree, and is of very little estimation in our Art, although by his swift flight he has elevated himself into the highest pitch above all the Luminaries, yet at the clipping of his wings, he must be reduced to the lowest light of all , and by corruption must be brought to a melioration, whereby the black must be changed into white, and the white into red; and the other Planets must pass through all the colors in the world, until they come to the proper super-abounding tincture of the triumphant King.

And so I tell you, that although *Saturn* be esteemed the meanest in the whole world, yet has he in him that power and efficacy, that if his pure Essence, which is beyond measure insensibly cold, be added to a currant fiery Metalline body, its running quality may be taken away, and nay be made a malleable body, as *Saturn* itself is, but of far greater fixity, which Transmutation has its original beginning and end from Mercury, Sulphur and Salt. This seems difficult to be understood by many, as indeed it is; but because the matter is vile, therefore must the intellect be acute and high, for there must be unequal states in this world to discern Masters from Servants.

From *Saturn* proceed many colors, that are made by preparation and Art, as black, ash-color, white, yellow, and red, and besides these in their mixtures arise other colors, so that the matter of the Philosophers must pass through many colors, before that great Stone can be exalted to its certain degree of perfection: For as often as a new passage is opened to the Fire, so often is a new form and species of its vestments given it for a reward, until the poor Artist gets riches, and needs not to borrow of another.

When the Lady *Venus* possesses the Kingdom, and does rightly distribute the Offices according to the customs of the Kings Court, she appears in Magnificent Splendor, and *Musica* bears before her a specious Ensign of a red color, wherein is painted *Charity*, very beautiful in her green robes, and in her Court *Saturn* is the Master President, who when he performs his office, *Astronomy* carries a black Ensign before him, whereon *Fides* is beheld painted in a yellow and red garment: *Jupiter* with his Scepter takes upon him the office of a Marshall, and before him goes *Rhetorica* with an Ensign of an ash-color, wherein *Spes* is most beautifully painted with splendid colors.

Mars is hardened in warlike affairs, and bears rule in the fiery heat, and *Geometria* draws before him a curtain of a bloody color, whereon Fortitude is discerned, clothed with a red garment: *Mercury* takes his place as Chancellor, and before him *Arithmetica* bears an Ensign of all colors, wherein Temperance is painted in glorious colors.

Sol he is Vice-Roy of the Kingdom, and *Gramatica* bears a yellow Ensign before him, whereon Justice is painted in a golden robe: Which Vice-Roy, although he has the greater power in his Kingdom, yet Queen *Venus* has blinded and conquered him, with her transcendent splendor.

Then *Luna* also appears, and *Dialectica* carries before her a Silver colored and shining Vail, whereon Prudence is painted in an azure color: And because *Luna's* husband died she painted the Office herself, lest Queen *Venus* should get into the government again; for she called her to an account of her office, then the Chancellor assists her, that a new Government may be established, and both then rule above the Queen: Understand, that one Planet must drive out and dispossess another of his government, office, possession, and power, until the best of all attain the highest power, and with the best and most fixed color given them by their first Mother, out of an innate constancy, love, and amity they obtain the victory: For the old world passes away, and the new is come in its place, and one Planet destroys another Spiritually, that he that is strongest continues till the last by feeding upon the other, two or three being overcome by only one.

For a final conclusion, you may understand hereby you must take *Cœlestial, Libra, Aries, Taurus, Cancer, Scorpio*, and *Capricorn*, and at the other end of the balance put *Gemini, Sagittarius, Aquarius, Pisces*, and *Virgo*; then cause that the golden *Lion*, leap into the lap of *Virgo*, so will that part of the scale be the weightiest, and weigh down the other; then let the twelve Signs of the Heaven come in opposition to the *Pleiades*. And so, after the finishing of all the colors of the world, there will at last be a conjunction and union, that the greatest comes to be the least, and the least to be the greatest.

If that the nature of the whole world remained,
Only in one state, form, or quality,
And other forms could not by Art be gained,
The wonders of the world would cease to be.
And Natures mysteries would not be raised,
For whose discoveries let God be praised.

The X. Key.

In our Stone made by me, and others long before me, are all the Elements, and all the Mineral and Metalline forms, yea, and all the qualities and properties of the whole world contained; for therein is found the greatest and strongest heat: For by its great internal fire the cold body of *Saturn* is warmed, and by that heating is changed into the best gold: In it also is found the greatest cold, for by its conjunction the hot Nature of *Venus* is temperated, and quick Mercury coagulated, and by the same reason, by its fixity is transmuted into the best fixed Gold: Because all these properties of our matter of the great Stone are infused by Nature, which properties are concocted and maturated by the degrees of fire, until they have attained the highest perfection, which cannot be done before that Mount *Etna* in *Sicilia* be consumed by its flames, and no more cold be found any more in the highest *Hyperborean* Mountains, which place may also be called *Filicium*.

If fruits are gathered before they are ripe, they are untimely and unprofitable, neither are they fit for use; so unless the Potter burn and concoct his wares enough in the fire, they are unfit for use, because they were not sufficiently Maturated in the fire.

So also concerning our *Elixir* you must diligently consider that a just time be given it, and that before that time nothing of its virtue be detracted, lest it be aspersed and esteemed for an unworthy thing.

It is evidently known that if the blossoms be plucked off, no fruit can grow there, therefore has it not for our Magistery, therefore he that makes too much haste seldom does good in our Art, but by haste more is spoiled then performed.

Wherefore let no searcher of Truth suffer himself to be deceived with overmuch desire to gather and pluck it before its time, lest the Apple slip from him, and the Stalk only remain in his hand; For in truth, if our Stone be not sufficiently maturated, no ripeness can be produced from it.

The Matter is dissolved in *Balneo*, and united by putrefaction in Ashes it produces flowers; in Sand all its superfluous humidity is dried away, but a quick fire maturates it with fixation, not that you must needs use *Balneum Mariæ, Fimus Equinus, Ashes* and *Sand* successively, but that the degrees and regiment of the Fire be so performed: For the Stone is made in an empty Furnace, of a threefold defense [or wall] firmly locked up, enclosed and concocted with a continual Fire, until all Clouds and Vapors vanish, and the Garment of Honor appear in the greatest splendor, and remains in one place of the lowest Heaven, and it be stopped in its course. And when the King can lift up his Arms no longer, he has obtained the government of the whole World, for he is made the King of everlasting fixity, no danger can ever hurt him, for he is become invincible: Now let me tell you, when your Earth is dissolved in its proper Water, dry away the Water thoroughly by its due Fire, then will the Air breathe into it a new life, and when the life is incorporated, you have a Matter, which deservedly can have no other name, than the great Stone of the World; that as a Spirit penetrates human and metalline bodies, and is the universal Medicine, without defect, for it expels the evil and preserves the good, it is also a melioration, to concoct the evil with the good: Its color declines from a shining redness to a purple, from a Ruby to a Gravate color, and in weight it is exceeding and very great.

Whosoever shall be adopted to this Stone, let him return thanks to the Creator of every creature, for that Cœlestial Balsome; and let him pray that for himself and his neighbor he may use it for the sustentation of this temporal life, and that he may enjoy eternal happiness in this valley of miseries, and in the other world. to come.

Let God be highly praised for his inexpressible gift and grace forever, AMEN.

The XI. Key

The eleventh Key of the multiplication of our Great Stone I will discover and reveal unto you by way of Parable after this Manner.

In the East there dwelt a Knight, called *Orpheus*, who mightily abounded in wealth, and did excel in all good things: He chose and took for his Wife his own sister *Euridice*: But when he could have no issue by her, he imputed it to his sins in choosing his own Sister for his wife: With his daily prayers he besought and begged the most high God, that he would communicate to him his Grace, and give way to his request.

Being sometime overcome with deep sleep, there came to him a man flying, named *Phoebus*, he touched his feet, which were very hot, and said, Most Noble *Hero*, you have travelled through many Kingdoms and Provinces, and many Towns, and Regions, and have undergone many dangers in the vast Ocean, and have sustained so much of the war, that you have acquired that Noble Order, and have merited that dignity before any other, having broken many weapons in Duels and Tournaments, and have often obtained honor by the Venerable Matrons: Therefore my Father in Heaven commanded me, that I should declare unto you, that your supplications were heard; therefore you are to take the blood of your right side, and the blood out of your Wife's left side, and the blood which was concealed in the heart of your Father and Mother, they are naturally two, and yet but one blood; conjoin these together, and cause again that they enter the Globe of the seven Wise Masters, nakedly enclosed; then is that mighty generation nourished with his own flesh, and is renewed with his own honorable blood; if you have done this rightly you shall leave a numerous generation, and issue begotten of your own body: But know, that the last seed in the eighth Revolution of time, will finish its course, as the first seed out of which at first it was made: If you do this often, and always begin anew, you shall see your Children's Children: That the great World shall be thoroughly replenished by the generation of the lesser, that may abundantly possess the Cœlestial Kingdom of the Creator.

This being ended, *Phoebus* fled away again, and the Knight awakened out of his sleep, and arose from his bed, and having done all things, as he was commanded, he not only found good success in his undertakings, but God also gave to him and his Wife many Children, who by their Fathers Testament did possess a memorable name, and the Honor of that Noble dignity did forever endure in that family with great riches.

Now, Son of Art, if you have understanding, you need no other interpretation; but if you have no understanding impute it not unto me, but to your own ignorance.

For I am prohibited to open this Lock any more, and I must obey, and observe its method; but to whom the Omnipotent gives to know, to him it is evident enough, and clearly written; yea, and more clearly than can be believed. I have described the whole process figuratively, and after the Philosophic manner, and as my predecessors have done, yea, and more plainly than them, for I have concealed nothing: If you remove the Veil from your eyes, you shall find that which many have sought, and few find, for the Matter is absolutely expressed by its Name: The beginning, middle, and end is also demonstrated.

The XII. Key.

A fencer who knows not how to use his weapon, it can be of no advantage to him, because he has not rightly learned the use thereof; another that better understands it than himself, fighting with him, the unskillful must needs be beaten by him; He that has well attained the Mastery of the Fencing School wins the Prize.

So he that has by the Grace of the Omnipotent God obtained the Tincture, and knows how to use it; so it happens unto him as was said of the Fencer, that knew not the use of his weapon: But seeing this twelfth and last Key is for the finishing of my Book, I will not detain you any longer in parabolical or figurative expressions, but without any obscurity I will discover this key of the Tincture in a most perfect and true process; Therefore observe my doctrine as it follows.

When the Medicine and Stone of the Philosophers is made, and perfectly prepared out of the true Virgins Milk, take thereof one part, of the best and purest Gold, melted and purged by Antimony, three parts, and reduce it into as thin plates as possibly you can, put these together into a Crucible, wherein you use to melt Metals, first give a gentle Fire for twelve hours, then let it stand three days and nights continually in a melting Fire, then are the pure Gold and the Stone made a mere Medicine, of a subtle, spiritual, and penetrating quality. For without the ferment of

Gold the Stone cannot operate, or exercise its tinging quality, being too subtle and penetrative; but being fermented and united with its like ferment, the prepared tincture obtains all ingress in operating upon other bodies. Then take of the prepared ferment one part, to a thousand parts of melted Metal, if you would tinge it, then know for a very certain truth, that it shall be transmuted into good and fixed Gold: For the one body embraces the other although they be not alike, yet by the force and power added to it, is made like it, like having its original from like.

He that uses this means, to him is revealed all truth. The Porches of the Palace have their goings forth at the end, and this Policy is not to be compared to any Creature; for it possesses *All* in *All*, as naturally and originally in this world can possibly be done under the Sun.

O Beginning of the first Beginning, consider the end.
O End of the last End, see to the Beginning.

And let the Middle be faithfully remembered by you, then will God the Father, Son, and Holy Ghost, give unto you, whatsoever you require for Spirit, Soul, and Body.

Of the First Matter of the Philosophers Stone.

A stone is found which is esteemed vile,
From which is drawn a fire volatile.
Whereof our noble Stone itself is made,
Composed of white and red that ne're will fade.
It's called a Stone, and yet is no Stone;
And in that Stone Dame Nature works alone.
The fountain that from thence did sometime flow,
His fixed Father drowned has also.
His life and body are both devoured,
Until at last his Soul to him restored:
And his volatile Mother is made one,
And alike with him in his own Kingdom.
 Himself also virtue and power has gained.
And far greater strength then before attained.
In old age also does the Son excel,
His own Mother, who is made volatile,
By Vulcans Art, but first its thus indeed,
The Father from the Spirit must proceed.
Body, Soul, Spirit, are in two contained,
The total Art may well from them be gained.
It comes from one, and is one only thing,
The volatile and fixt, together bring.
It is two and three, and yet only one.
If this you do not conceive, you get none.
Adam in a *Balneo* resides
Where *Venus* like himself abides,
Which was prepared at the old *Dragods* cost,
Where he his greatest strength and power lost.
Its nothing else said one Philosopher,
But a *Mercurius Duplicatus*.
I will say no more, its name I have shown,
Thrice happy is the man to whom its known.
Seek for it there, and spare not cost and pains,
The end will crown the work with health and gains.

A Brief Appendix

And plain Repetition or Reiteration of Basilius Valentinus The Monk of the Order of St. Bennet, to his Book of the Great Stone of the Ancients.

I, Basilius Valentinus, Monk of the Order of Saint Bennet, have written a small Treatise, and as the Ancients, have revealed in a Philosophic manner, how that most excellent treasure may be attained, by which the true Philosophers did exceedingly prolong their lives.

And although, as my conscience bears me witness before the Highest in the Heavens, to whom all secrets are manifest, I have written no untruths, but have made the truth itself so plain that understanding men need no more light. (For my Theory written for then, which was confirmed and made plain by the practice of the twelve Keys, is sufficient.) Nevertheless, the unquietness of my mind so wrought with me, through my various cogitations, that I undertook to add this small Tract, to demonstrate by a shorter way, and as it were by this means to purify that burning light, whereby every lover of true wisdom, may have his desire the more fulfilled by that splendor and clearness. And although many esteem it clear enough already, and so heap upon me the burden of many evils, yet everyone know, that to these that are dull of understanding, they will find that which they seek to be difficult enough, but to the adepted plain and easy; therefore my searcher of truth attend to my Instructions, and you shall find the true way to Art.

For I have written nothing but what I shall bear witness unto after my death, and at the Resurrection of my body.

You shall faithfully and truly find the shorter way, in the following discourse, for my sayings are founded in simplicity, and not in Sophistical Expressions.

I have mentioned and demonstrated, that all things are made and compounded of three Essences, viz., of Mercury, Sulphur, and Salt; and it is true what I have said.

But know this, that the Stone is made of one, two, three, four, and five: Of five, that is, the quintessence of its Matter; Of four, are understood the four Elements; Of three, they are the three principals of all things; Of two, for that is the double Mercurial substance; of One, that is the *Ens primum* of all things, which flowed from the *Fiat* of the first Creation.

Many well minded Artists may be doubtful by all these sayings, to attain the foundation and understanding of the following discoveries, therefore I shall first very briefly speak, of Mercury; secondly, of Sulphur; thirdly, of Salt; for these are Essences of our Matter of the Stone.

First know that no common Argent vive is fit for our use; but our Argent vive is made of the best Metal by the Spagyric Art, pure, subtle, clear, splendent, as a Fountain, transparent as Chrystal, without any impurity; of this make a Water or incombustible Oil; for Mercury was at the first water, as all Philosophers agree to this my saying and doctrine.

In this Mercurial Oil, dissolve its proper Mercury, out of which the Water was made, and precipitate that Mercury with its proper Oil, then have you a double Mercurial substance; and know that your Gold must first be dissolved in a certain Water, expressed in my second Key, after its purification, as in the first Key, and must be reduced into a subtle Calx, as is mentioned in the fourth Key; and then the said Calx must be sublimed by Spirit of Salt, and precipitated again, and by reverberation reduced into a subtle Powder; then its own proper Sulphur will the more easily enter into its own substance, and be in amity with it, for they wonderfully love each other. So have you two substances in one, and is called the Mercury of the Philosophers, and yet is but one substance, that is, the first ferment.

Now follows what is to be said of Sulphur.

Your Sulphur you must seek in the like Metal, then you must know how to extract it out of the body of the Metal by purification, and destruction of its form and reverberation, without any corrosive, whereof I gave you a hint, and minded you of it also in the third Key: Then dissolve the Sulphur in its own proper blood, whereof it was made before its fixation, according to its due weight shown in the sixth Key; then have you nourished and dissolved the true Lion, with the blood of the green Lion; for the fixed blood of the red Lion, is made out of the volatile blood of the green Lion; therefore are they of one nature. And the volatile blood makes the fixed blood volatile, and the fixed likewise makes the volatile blood fixed, as it was before its solution. Then set them together in a gentle heat, until the whole Sulphur be dissolved; then have you the second ferment, nourishing the fixed Sulphur with the volatile, as all Philosophers agree with me herein: this afterwards is driven over with Spirit of Wine, red as blood, and is called *Aurum Potable*, whereof there is no reduction to a body.

I will also give you my opinion of the Salt of the philosophers.

Salt makes fixed and volatile, according as in its degree it is ordered and prepared: For the Spirit of Salt of Tartar, if it be drawn *per se*, and without addition, makes all metals volatile by resolution and putrefaction; and resolves them into a true Vive, or current Mercury, as my practical doctrine holds forth.

Salt of Tartar *per se* fixes most firmly, especially if the heat of *Calx Vive* be incorporated with it, for both these have a singular degree of fixing.

So also the vegetable Salt of Wine both fixes and makes volatile according to the divers preparation thereof, as its use requires, which certainly is a great mystery of Nature, and a wonder of the Philosophic Art.

If a man drink Wine, and out of his Urine a clear Salt be made, that is volatile, and makes other fixed things volatile, and carries it over the helm with it, but it fixes not; and although the Man drink nothing but wine, out of whose urine the Salt was made, yet it has another property, than the Salt of Tartar, or of the Feces of Wine: For there is made a transmutation in the body of Man, so that out of a Vegetable, that is, out of Spirit of Wine, an animal Spirit of Salt is made, Horses by the corroboration of their natural virtue, do transmute, oats, hay, and such like, and convert it into fat and flesh; so do the Bees make Honey out of the best of Flowers and Herbs.

So, understand of other things: This Key and Cause consists only in putrefaction, from whence such a separation and transmutation takes its original.

The spirit of common Salt, which is drawn after a peculiar manner, makes Gold and Silver volatile, if a small quantity of the spirit of the *Dragon* be added to it, it dissolves it, and carries it over with it *per Alembicum*, as also does the *Eagle* with the *Dragons* Spirit, which dwells in stony places; but if anything be melted with Salt, before the Spirit be separated from its body, it fixes much more than it volatilizes.

This I further tell you, if the spirit of common Salt be united with spirit of Wine, and both be three times distilled over together, then it waxes sweet, and loses its acrimony: This prepared spirit does not corporally dissolve Gold, but if it be poured on a prepared Calx of Gold, it extracts its highest tincture and redness; which if it be rightly done, it reduces pure and white *Luna* into the same color whereof its body was, before it was extracted: Also the old body will again attain its color by the Love of enticing *Venus*, being descended from the same original, state, and blood, whereof this is not a place to speak any further.

I know also the Spirit of Salt destroys *Luna*, and reduces it into a Spiritual Essence, according to my instruction, from whence afterwards *Luna potable* may be prepared which Spirit of *Luna* is appropriated to the Spirit of *Sol*, as Man and Wife, by the copulation and conjunction of the Spirit of Mercury, or its Oil.

The spirit lies in Mercury, seek the tincture in Sulphur, and the coagulation in Salt, then have you three matters, which may again produce some perfect thing, that is, the Spirit in Gold fermented with its own proper Oil. Sulphur is plentifully found in the propriety of most precious *Venus*, which inflames the fixed blood gotten of her: The Spirit of the Philosophic Salt gives victory to coagulation, although the Spirit of Tartar, and Spirit of Urine, together with the true Acelum may do much: For the Spirit of Vinegar is cold, and the Spirit of Calx Vive is very hot, therefore are they esteemed and found to be of contrary natures: Now I speak not according to the Philosophic custom: But it does not become me to discover more plainly, how the inner doors are locked.

This I faithfully tell you for a farewell: Seek your matter in a Metalline substance, make thereof Mercury; which Ferment, with Mercury: Then a Sulphur, which Ferment with its proper Sulphur, and with Salt reduce it into Order, distill them together, conjoin then all according to their due proportion, then will it become that one thing, which before came from one; coagulate and fix it by a continual heat; then multiply and ferment it three times, according to the doctrine of my two last keys, then shall you attain and find the end and conclusion of your desire: The use of the tincture, the twelfth key, has absolutely the certain process, without any doubtful expressions.

THANKS BE UNTO GOD

For a Conclusion of this Appendix, I must needs tell you that out of black *Saturn* and friendly *Jove*, a Spirit may be extracted, which is afterwards reduced into a sweet Oil, as its noblest part, which Medicine, *particulariter* does most absolutely take away the nimble running quality from common *Mercury* and brings him to a melioration, as I taught you before.

AN ADDITION

Having thus attained the Matter, nothing remains but that you look well to the Fire, that you observe its Regiment, for herein is the highest concernment, and the end of the work: For our Fire is a common Fire, and our Furnace is a common Furnace, although they that were before me have written, that our fire is not a common Fire, yet I tell you in truth, that they did after their manner conceal all Mysteries, because the Matter is vile, and the Work is but little, which the Regiment of the fire only furthered and manifested.

The Fire of the Lamp with Spirit of Wine is unprofitable, the expense thereof would be incredible; *Fimus Equius* spoils it, for it cannot perfect the work by the right degrees of Fire.

Many and various Furnaces are not convenient, for in our threefold Furnace only the degrees of fire are proportionally observed; therefore, let no pratling Sophister lead you into errors with many furnaces: And as our Furnace is common, so is our Fire common, and as our Matter is common, so is our Glass likened to the Globe of the earth: You need no further instructions concerning the Fire, its Regiment, or Furnace.

For he that has the Matter will soon find a Furnace, he that has Meal can soon find an Oven, and needs take no further care for baking of the bread.

There is no further need to write peculiar books of this subject, only observe the Regiment of the Fire, to know how to distinguish between cold and hot, if you attain this you have done the work, and brought the Art to a conclusion: For which let the Creator of Nature be praised forever, AMEN.

Of Mercury

There are several sorts of *Mercury*. Mercury of Animal and Vegetables is merely a fume of an incomprehensible being, unless it be caught, and reduced to an oil, then is it for use. But Mercury of Metals is of another condition, as that also of Minerals: Though the same also may be compared with a fume, yet is it comprehensible and running. One Mercury is better and nobler than the other: For the *Solar* Mercury is the best of them all; next unto that is the *Lunar* Mercury, and so forth. There is a difference also among Salts and Sulphurs: Among the Mineral Salts, that carries away the Bell, which is made of Antimony: And that Sulphur, which is drawn from Vitriol, is preferred before all others. Mercury of Metals is hot and dry, cold and moist, it contains the four qualities.

There are Medicaments prepared of it, of a wonderful efficacy, of several sorts and forms, which is the reason, why there is such a variety of virtues therein: In Mercury lies hid the highest *arcanum* for man's health, but is not to be used crude, but must first be prepared into its essence. It is sublimed with Copper-water, and is further reduced into an Oil. There is an Oil made of it *per se*, without any corrosiveness, which is pleasant and fragrant: Several sorts of Oils with additionals can be made of it, good for many things. It is prepared also with Gold, being first made into an *amalgame*; there is made a precipitate of it in water, wherein it dissolves green, like unto a smarag'd, or Crysolith: The volatile Mercury serves for outward use, if a separation is made by some means, and is brought into subtle, clear *liquor*, and then to a red-brown powder, and its received corrosiveness is separated, then it may do well for other uses.

The mixed Mercury serves for inward use.

Mercury being purged, is precipitated with the blood of *Venus*, is well digested with distilled Vinegar, and thus his corroding quality is taken off: Have a care what quantity you minister, if it being given in a true dose, then it does its part very well: But for its operation, it is not equally sublimed unto the fixed, its coagulation is found in Saturn, his malleableness is apparent, when he is robbed of his life: he contains his own Tincture upon white and red, being brought in his fixed coagulation unto a white body, is tinged again by Vitriol water, and being reduced unto Gold, is graduated by Antimony. Though that blood-thirsty *Iron Captain* with his spear assaults *Mercury* very much, yet he alone cannot conquer him, unless cold *Saturn* come in to hide him, and *Jupiter* command the peace with his Scepter. Such processes being finished, then the Angel *Gabriel*, the strength of the Lord, and *Uriel* the light of God has shown mercy unto humble *Michael*, then *Raphael* can make use of the highest medicine, nothing can prevail against the Medicine. Thus much be spoken of *Mercury*: now I swing myself from hence, and fly to a place where frost and heat can better be tolerated, and endured.

Of Antimony

It falls very difficult to Mechanics, to have done learning with their compasses: because that great *Architect JEHOVAH* has reserved many things for his own power. In the same condition we find *Antimony*, it is very difficult to find out all the mysteries that are hid therein; its virtue is miraculous, its power is great, its color hidden therein; is various, its crude body is poisonous, yet its essence is an antidote against poison, is like unto Quick-silver, which ignorant Physicians can neither comprehend, nor find, but the knowing Physician, believing it to be true, as having made many experiments with it.

This Mineral contains much of Mercury, much of Sulphur, and little of Salt, which is the cause why it is so brittle and appliable: for there is no malleableness in it, by reason of the small quantity of Salt, the most amity it bears unto *Saturn* is by reason of *Mercury*: For Philosophers Lead is made out of it, and is affected unto Gold, by reason of its Sulphur: For it purges Gold, leaving no impurity in it; there is an equal operation in it with Gold, if well prepared, and ministered to man Medicinally: it flies out of the fire, and keeps firmly in the fire, if it be prepared accordingly. Its volatile spirit is poisonous, purges grievously not without damage unto the body; its remaining fixedness purges also, but not in that manner, as the former did, provokes not to stool, but seeks merely the disease, where ever it is, penetrating all the body, and the members thereof, suffers no evil to abide there, expels it, and brings the body to a better condition.

In brief, *Antimony* is the Lord in Medicinals, there is made of it a *Regulus* out of Tartar and Salt, if at the melting of Antimony some Iron filings be added, by a Manual used, there comes forth a wonderful Star, which Philosophers before me, called the *Signet-Star*. This Star being several times me]ted with cold *Earth-Salt*, it grows then yellowish, is of a fiery quality, and of a wonderful efficacy: This Salt afterwards affords a *liquor*, which further is brought to a fixed incombustible Oil, which serves for several uses.

Besides, there are made of common *Regulus* of Antimony curious flowers, either red, yellow, or white, according as the fire has been governed. These flowers being extracted, and the extract, without any addition *per se* being driven into an Oil, have an admirable efficacy. This extraction may be made also with Vinegar of crude Antimony, or of its *Regulus*; but it requires a longer time, neither is it so good as the former preparation.

And being reduced into a *Phitistæa*, there is a glass made of it *per se*, of which I made mention in my eighth Key, which is extracted also, then abstracted there remains a powder of incredible operation, which may safely be used, after it has been edulcorated. This powder, being dissolved, heals wounds, sores, &c. causing no pains: This powder being extracted once more with Spirit of Wine, or driven through the Helmet, with some other matter, affords a sweet Oil; to speak further of it is needless.

Antimony is melted also with cold *Earth-Salt*, dissolved, and digested for a time in spirit of wine, it affords a white, fixed powder, is effectual against *morbus Gallicus*, breaks inward Impostums; it has several virtues besides. You must learn to prepare Antimony yourself, lay hands on, dive into its inward qualities, you will meet with wonderful matters: for my conscience will not suffer me to discover all its qualities: I desire not to load the Physicians curses upon me, which were at great expenses, and toiled much in burning of Coals about its preparation, if I should rob them of their lively-hood. Therefore, learn this also, as your predecessors did; seek as I have done; then you will find also, what others have told of.

There is made an Oil also of Antimony, the flying Dragon being added thereunto, which being rectified thrice, then it is prepared: though a *Cancer* were never so bad, and the *Wolf* never so biting, yet these with all their fellows, be they *Fistula's*, or old Ulcers, must fly and be gone: The little powder of the flying Dragon prepared with the Lions blood, must be ministered also, three, or four Grains for a Dose, according to the parties age and complexion.

A further process may be made with this Oil with the addition of a water, made of Stone Serpents, and other necessary Spices; not those which are transported from the Indies: this powder is of that efficacy, that it radically cures many Chronical diseases.

There is made a red Oil of Antimony, Calx vive, Sal-armoniac, and common Sulphur, which has done great cures in old ulcers: with stone Salt, or with common Salt, there is forced from Antimony a red Oil, which is admirably good for outward Symptoms.

There is made a sublimate of Antimony, with Spirit of Tartar and Salmiac, being digested for a time, which, by means of *Mars* is turned into quick Mercury. This Antimonial Mercury has been sought of many, but few have gotten it: Which is the reason why its praise is not divulged, much less is its operative quantity known: If you know how to precipitate it well, then your Arrow will hit the mark, to perform strange matters; its qualities ought not to be made common.

It is needless to describe its combustible Sulphur, how that is made of Antimony, it is easy and known: But that which is fixed, is a secret, and hidden from many. If an Oil be made of it, in which its own Sulphur is dissolved, and these be fixed together, then you have a Medicine of rare qualities, in virtue, operation, and ability far beyond Vegetables.

Quick-silver being imbibed with quick Sulphur, melted with Antimony for some hours in a Wind-oven, the Salt of the remainder being extracted with distilled Vinegar, then you have the *Philosophers Salt*, which cures all manner of Agues.

There is an *acetum* made of Antimony, of an acidity, as other *acetums* are; if its own Salt be dissolved in the *acetum*, and distilled over, then this *acetum* is sharpened, which is an excellent cooler in hot swellings, and other inflamed symptoms about wounds, especially if there be made an unguent of it together with *anima* of Saturn.

The Quintessence of Antimony is the highest Medicine, the noblest and subtlest found in it, and is the fourth part of a *Universal* Medicine. Let the preparation of it be still a mystery, its quantity, of Dose is three grains, there belong four instruments to the making of it, the furnace is the fifth, in which *Vulcan* dwells, the Manuals, and the government of fire afford the ordering of it.

You Physicians, if you be wise, seek out this Medicine in that subject, where it lies in, and may be found best, and most effectual. I forbear to speak further of Antimony, let *Justinian* judge of the rest.

Of Copper-water.

If I could prevail with *Apollo* to be merciful, and to give liberty to his *Muse* to be my assistant in the describing of Art and Wisdom, then would I bring in an offering unto *Minerva*, whereby the Gods of Wisdom might take notice of a grateful mind for their gifts they had bestowed; and I would write of a mineral, whose Salt is set forth in the highest manner, whose great and good qualities are of that transcendency, that reason is not able to comprehend, or to conceive of them. It went generally by the name of *Copper-water*, to make the meaning and sense of it plain; let men know, and be thus informed of it, that *Vitriol* contains two spirits, a white, and a red one: The white spirit is the white Sulphur upon white, the red spirit is the red Sulphur upon red; He that has ears let him hear!

Observe diligently, and remember every word, for they are of a large extent, every word is as ponderous as a Center stone. The white spirit is sour, causes an appetite, and a good digestion in a man's stomach. The red spirit is yet sourer, and is more ponderous than the white, in its distilling a longer fire must be continued, because it is fixed in its degree. Of the white by distilling of Sulphur of *Lune* is made *argentum potable*. In like manner the Gold, being destroyed in the spirit of common Salt, and made Spiritual by distilling, and its Sulphur taken from it, and joined with a red spirit in a due Dose, then it may be dissolved, and then for a time putrefied in spirit of Wine, to be further digested, and often abstracted, that nothing remain in the bottom, then you have made an *aurum potable*, of which great volumes have been written, but very few of their processes were right. Note, that the red spirit must be rectified from its acidity, and be brought into a sweetness, subtly penetrating of a pleasant taste and sweet fragrancy.

I have told you now great matters, which slipped from me against my intention, the sweet Spirit is made of Sulphur of Vitriol, which is combustible, like other Sulphur, before it is destroyed: For the Sulphur of Philosophers, (note it well) is not combustible; its preparation needs not to be set down, being easy, requiring no great pains nor great expense, to get a combustible Sulphur out of vitriol.

This sweet Oil is the essence of Vitriol, and is such a Medicine, which is worthy the name of the third Pillar of the Universal Medicine. The Salt is drawn from *Colchotar*, and is dissolved in the red, or white Oil, or in both, and is distilled again, if it be fermented with *Venus*, it performs its office very well: for it affords such a Medicine, which at the melting tinges pure Iron into pure Copper.

Colchotar of Sulphur affords true fundamentals unto healing of perished wounds, which otherwise are hardly brought to any healing: and such sores, which by reason of a long continued white redness will admit of no healing, *Colchotar* affords an ingress thereunto, setting a new foundation; that quality and virtue is not in the *Colchotar*, but the spirit together with the Salt are the Masters, which dwell therein.

There is made of Copper and Verdigreece a Vitriol of a high degree, and is far spread in its tincture: There is a Vitriol made of Iron also, which is of a strange quality: For Iron and Copper are very nigh kind one to another, belong together, as man and wife, this mystery should have been concealed, but being it is of great concernment, I could not forbear but to speak of it.

Vitriol is corroded with Sal-armoniac, in its sublimation there arises a combustible Sulphur, together with its Mercury, of which there is but little, because it has most of Sulphur. If the same Sulphur be set at liberty again by the *Eagle*, with spirit of Wine, there can be made a Medicine of it, as I told you formerly of it. Though there be a nearer way to make a combustible Sulphur out of Vitriol, as of its precipitation upon a precedent dissolution, by the Salt, or liquor of Tartar, as also by a common *lixivium* made of Beech-ashes; yet this is the best reason, because the body of Vitriol is better, and more opened with the Key of the *Eagle*. There are other mysteries hidden in Vitriol, which in your operative quality are excellent, and are known apparently, as *Venus* and *Mars* bear real record in their spirits, the same does knowledge *Sol* and *Lune*: But I do not intend at this time to write a perfect book of *Chirurgery*: and to make relations of particulars, in commendation of Vitriol: I have already written too much of it, you are to learn and search also; you will find that Vitriol needs no Proctor to speak for it, and it will sufficiently inform you of an absolute Chirurgic book, contained in its nature as a third part of the *universal*, against all manner of diseases.

In the closing hereof I tell you this much, that there is not found in its nature, neither cold or moist quality, but is of a hot and dry substantial quality, and is the reason, why by its super-abounding calidity it heats other things, digests them, and at last it brings them to a full maturity, the fire being continued for a certain time.

The things I write of Vitriol, I have not begged or borrowed from other men's writings, but found them so in my long-continued practice, whereby nature enabled me to become a *Soothsayer*, by permission of the highest Creator, that that nobly implanted quality might be avouched by a faithful and true evidence of one of her devoted Disciples.

And I speak thus much for a *memorandum*, that if *Paris* can keep safely *Helena* without troubles, that the noble City of *Troja* in *Greece* be no more ruined and destroyed, and *Priamus* together with *Menelaus* be no more afflicted and distracted thereby, then *Hector* and *Achilles* will agree well enough, to obtain that royal Race, without going to war for it, and be Possessors of that Monarchy in their Children's Children, and their off-spring and posterity for the enlarging of their Dominions, by increasing their riches infinitely, against which no enemy dares stir.

Of Common Sulphur

The usual common Sulphur is not so perfectly exalted in its degree, and brought unto maturity, as it is found in Antimony and Vitriol. There is made of it *per se*, an Oil against putrid stinking wounds, destroying and killing such worms, which grow in them; especially if that little Salt in it be dissolved from its Sulphur.

There is made of it a Balsam with Sallet Oil, or Oil of Juniper, in like manner with the white spirit of Terpentine, and is of a red color, is made thus: take flowers of Sulphur, made with the *Colchotar* of Vitriol, digest them for a time in horse-dung, or any other way, this Balsam may safely be used for such, that are in a Consumption of the Lungs, especially if rectified several times with spirit of Wine, drawn-over, and separated, that it be blood red. This Balsam is a preservative against corruption and rottenness.

The Quintessence of Sulphur is in a Mineral, where a Sulphureous flint is generated: these beaten pebbles being put in a glass, and on it be poured a strong Aquafort, made of Vitriol and Saltpeter, and let dissolve what may be dissolved, abstract that water, the remainder must be well dulcified, and reverberated to a redness, pour on that spirit of Wine, extract its tincture, afterwards circulate for a time in the Pelican, let all the essence of Sulphur be separated, it stays below the spirit of Wine, like fat Sallet Oil, by reason of its ponderousness: its Dose of six

Grains is found to work sufficiently. If you digest in this essence of Sulphur, Myrrhe, Aloes, and other Spices, it extracts their virtues, and makes it into a Balsam, which suffers no flesh, or other parts that are subject unto putrefaction, to fall into rottenness, for which cause the Ancients have put the name to it: *Balsamus mortuorum.*

Thus I close to speak any further of combustible Sulphur. There may be made an Oil of it, which is found very useful, the Sulphur must be sublimed in a high instrument with a good heat, which sublimation in a long-time changes into a liquor or Oil, standing in a humid place; but being I do not intend to use any prolixity of words, I let it rest so. There may be cocted a Liver out of common Sulphur, which is turned into milk: and it may also be changed into a red Oil, with linseed Oil; many other Medicinals may be made out of Sulphur Its flowers, essence, and Oil, are preferred before the rest, together with the white and red fixed Cinobar, which are made of it, because in them is found a mighty virtue.

Of Calx-vive.

The secrets of Quick-lime are known to few men, and few there are, which attained to a perfect knowledge of its qualities: but I tell to you a real truth, that though Lime is contemptible, yet there lie great matters therein, and requires an understanding Master, to take out of it what lies buried in it: I mean to expel its pure spirit, which collaterally stands in affinity with Minerals, is able to bind, and help to make fix the volatile spirits of Minerals: for it is of a fiery essence, heats, concocts, and brings unto maturity in short time, when in many years they could not be brought to it: the gross earthly body of it does not do the feat, but its Spirit does it, which is drawn out of it: this spirit is of that ability, that he binds and fixes other volatile spirits. For note, the spirit dissolves *Oculi Cancrorum*, dissolves crystals into a liquor: these two being duly brought into an unite *permodum distillationis* (I will say nothing at this time of Diamonds and such like stones) that water dissolves and breaks the stone in the bladder, and the Gouty Tartar settled into the joints of hands and feet, suffer not any Gout to take root in these parts, this rare secret I taught one of my faithful Disciples; and the great Chancellor of the invincible *Cæsar*, is still thankful unto me for it, and many great persons besides.

Quick-lime is strengthened, and made more fiery, and hot, by a pure unsophisticated spirit of Wine, which is often poured on it, and abstracted again, then the white Salt of Tartar must be ground with it, together with its additionals, which must be dead, and contain nothing, then you will draw a very hellish spirit, in which great mysteries lye hid. How this spirit is gotten, I told you, observe it, keep it, take it for a fare-well.

Of Arsenick.

Arsenic is in the kindred of Mercury and Antimony, as a Bastard in a Family may be: its whole substance is poisonous and volatile, even as the former two, in its external color to the eye, it is white, yellow, and red, but inwardly it is adorned with all manner of colors, like to its Metals, which it was fain to forsake, being forced thereunto by fire. It is sublimed *per se*, without addition, and also in its subliming there are added several other matters, as occasion requires. If it be sublimed with Salt and Mars then it looks like a transparent Crystal, but its poison stays still with it, unfit to be joined, or added to Metals, has very little efficacy to transmute any metal.

The subterranean Serpent binds it in the Union of fire, but cannot quite force it, that it might serve for a Medicine for man and beast, if it be further mixed, with the Salt of a vegetable Stone, which is with Tartar, and is made like unto an Oil, it is of great efficacy in wounds, which are of a hard healing: it can make a Coat for deceitful *Venus* to trim her handsomely, that the inconstancy of her false heart may be disclosed by her wavering servants, without gain, with her prejudice and damage.

When *Antimony* and *Mars* are made my companions, and am exalted by them to the top of *Olympus*, then I afford a Ruby in transparence and color to that, which comes from *Orient*, and am not to be esteemed less than it: if I am proved by affliction, then I fall off like a flower, which is cut off and withers: therefore nothing can be made of me, to fix any Metal, or tinge it to any profit, because I forsook my body totally, and distributed my Coat, to play, and lot to be cast for it: therefore let no man either praise or dispraise me, unless he have for very hunger taken a pound of me into the body; through it he gets an Antidote to save his life; however, he shall get nothing out of Metals by it: in other things he may have a Treasure in it, unto which few are comparable to it.

I Arsenic say of myself at the closing hereof, that it is a very difficult thing, to find my right and due preparation, my operation is felt exceedingly, if made trial of, and is a great danger, if ignorant men make use of me: he that can be without me, let him go to my kindred: and if you can equalize me with them, that I may share with them in the inheritance, then all the world shall acknowledge, that my descent is from their blood: but it is a very hard task for any man, to set a shepherd into a royal seat to make him King. But Patriarchs being descended from shepherds, and were preferred to royal dignities, I will therefore prescribe no limits, nor pass any judgement: For wrong and right may be found in this leaf.

However, take you notice, that I am a poisonous volatile bird, have forsaken my dearest, and most confiding friend, and separated myself as a Leper, which must live aloof off from other men. Cure me first of my infirmity, then I shall be able to heal those, which have need of me, that my praise may be confirmed by poison, and my name for an everlasting remembrance, to the honor of my Country, is nothing inferior unto *Marcus Curtius*, and it will be found in the end, in what manner *Hannibal* and *Scipio* were reconciled.

Of Saltpeter.

Two Elements are predominant in me, as fire and air, the lesser quantity is water and earth: I am fiery, burning, and volatile. There is in me a subtle spirit, I am altogether like unto Mercury, hot in the in-side, and cold in the out-side, am slippery and very nimble at the expelling of my enemies. My greatest enemy is common Sulphur, and yet is my best friend also, for being purged by him, and clarified in the fire, then am I able to allay all heats of the body, within and without, and am one of the best Medicaments, to expel and to keep off the poisonous plague.

I am a greater cooler outwardly than *Saturn*, but my spirit is more hot than any, I cool, and burn, according as men will make use of me, and according as I am prepared.

When Metals are to be broken, I must be a help, else no victory can be obtained: Be the understandings great or small. Before I am destroyed I am a mere Ice, but when I am anatomized, then am I a hellish fire. If *Pluto* can master *Cerberus*, to make him take his dwelling again in the isle of *Thule*, then he may snatch a piece of love from *Venus*, then *Mars* must submit, and may live richly with *Lune*, which may equally be exalted to the Crown of the honorable King, and be placed with him in equal honor and dignity.

If I shall happily enjoy my end, then my Soul must be driven out cunningly, then I do all what lies in my power, of myself alone I am able to effect nothing. But my love is to a jolly woman, if I am married unto her, and our copulation be kept in Hell, that we both do sweat well, then that which is subtle, flings away all filthiness, then we leave behind us rich Children, and in our dead bodies is found the best Treasure, which we bequeathed in our last Will and Testament.

Of Sal-armoniac.

Sal-armoniac is none of the meanest Keys, to open Metals thereby: therefore the Ancients have compared it with a volatile Bird, it must be prepared, else you can do no feats with it; for if it be not prepared, it does more hurt than good unto Metals, carries them away out at the Chimney-hole; it can elevate and sublime with its swift wings the tincture of Minerals, and of some Metals, to the very Mountains, where store of snow is found, usually even at the greatest heat of Summer, if it be sublimed, with common Salt, then it purges and clears, and may be used safely.

He that supposes to transmute Metals with this Salt, which is so volatile, surely he does not hit the nail on the head, for it has no such power: But to destroy Metals, and make them fit for transmutation, in that respect it has sufficient power: For no Metal can be transmuted, unless it be first prepared thereunto. My greater strength which lies in me, may be drawn from me by subliming and cementing. The greatest secret in me you will find, when I am united with *Hydra*, which is to devour and swallow me, that I also may turn with her to be a water Serpent, then have I prepared a Bath for the *Nympha*, and have gotten power to make ready a Crown for the King, that the same may be adorned with Jewels, and may with honor and glory be set on his head.

Of Tartar.

This Salt is not set down in the book of Minerals, but is generated of a vegetable seed, but its Creator has put such virtue into it, that it bears a wonderful love and friendship unto Metals, making them malleable: it purges *Lune* unto a whiteness, and incorporates into her such additionals, which are convenient for her, being digested for a time with Minerals, or Metals, and then sublimed and vilified, they all come unto quick Mercury, which to do, there is not any vegetable Salt beside it: is not this a wonderful thing! That Orator is yet to be born, which shall be of that ability and eloquence as to express sufficiently all the mysteries hid in it. But to make out of it the Philosophers Stone, is no such matter: Being it is a vegetable, and that power is not given to any of the vegetables. It is in *Medicina* a very good remedy, to be used inwardly and outwardly; its Salt being made Spiritual and sweet, it dissolves and breaks the stone in the bladder, and dissolves the coagulated Tartar of the Gout, settled in the joints, or anywhere besides. Its ordinary spirit, which is used for opening of Metals being used and applied outwardly, also lays a foundation for healing of such Ulcers, which admit hardly any healing, as there are Fistula's, Cancers, Wolves, and such like. I know nothing more to write of Tartar, for having separated itself, and left its noblest part in the Wine.

Of Vinegar.

In *Alchemy* and *Medicina*, nothing almost can be prepared, but Vinegar must set a helping hand to it. Therefore, I thought it convenient to let it have its due praise and commendation, especially to insert it here in this treatise. In *Alchemy* it is used to set Metals and Minerals into putrefaction. It is used also to extract their essences and tinctures, being first prepared thereunto, even as the spirit of Wine is usual to extract the tinctures from vegetables.

In Physick it deserves its praise also, for it takes the pure from impure, and is a *Separator*, and takes from the Mineral Medicaments their sharpness and corrosiveness, fixes that, which is volatile and is a great defendant against poison, as I told you, when I spoke of the Antimonial glass.

Vinegar is used inwardly also, and both men and beast are benefited thereby: Outwardly it is applied to hot inflammations and swellings, for a cooler. Spirit of Wine and Vinegar are of great use, both in Alchemy and Physick, both have descent from the Wine, are of one substance; but differ in the quality, by reason of putrefaction, the Vinegar got there, of the which I told you formerly.

I must acquaint you with one thing, which is this, that this is not the Philosophers Vinegar, our Vinegar, or *acetum* is another liquor, namely a matter itself: For the Stone of Philosophers is made out of *Azot* of Philosophers, which must be prepared with ordinary distilled *Azot*, with spirit of Wine, and with other waters besides, and must be reduced into a certain order.

Note this for a *memorandum*, if distilled pure Vinegar be poured upon destroyed *Saturn*, and is kept warm in *Marie's Bath*, it loses its acidity altogether, is as sweet as any Sugar, then abstract two, or three parts of that Vinegar, set it in a Cellar, then you will find white transparent stones, like unto Crystals, these are an excellent cooler and healer of all adult and inflamed symptoms. If these Crystals are reduced into a red Oil, and poured upon Mercury, precipitated by *Venus*, and proceeded in further as it ought; if that be his rightly, then neither *Sol*, nor *Lune* will hinder you from getting riches.

Of Wine.

The true vegetable stone is found in Wine, which is the noblest of all vegetables: it contains three sorts of Salt, three sorts of Mercury, and three sorts of Sulphur.

The first Salt sticks in the wood of the Wine, which if burnt to ashes, and a lixivium made of it to have its Salt drawn forth, which must be coagulated. This is the first Salt.

The second Salt is found in Tartar, if that be incinerated, then draw its Salt forth, dissolve and coagulate it several times, and let it be sufficiently clarified.

The third Salt is this, when the wine is distilled it leaves *feces* behind, which are made to powder, its Salt can be drawn out with warm Water. Each of these Salts has a special property: In their Center they stand in a harmony, because they descend from one root.

It has three sorts of Mercury, and three sorts of Sulphur. The first Oil is made of the Stem, the second Oil is made out of crude Tartar, the third is the Oil of Wine. There is a strange property in the spirit of Wine: For without it there cannot be extracted any true tincture of *Sol*, nor can there be made without it any true *aurum potable*: But few men know how a true spirit of Wine is made, much less can its property be found out wholly.

Several ways have been tried to draw, and to get the spirit of Wine without sophistication, as by several instruments and distillings with metalline Serpents, and other strange inventions, of Sponges, Papers, and the like. Some caused a rectified *aquavitæ* be frozen in the greatest frost, expecting the phlegm thereof should turn to Ice, the spirit thereof to keep liquid, but nothing was done to any purpose.

The true way for the getting of it, I told you of at the end of my manuals: For it must be subtle, penetrating, without any phlegm, pure, aerial, and volatile, so that air in a magnetic quality may attract it, therefore it had need to be kept close in: It is of a penetrating and effectual operation, and its use is several.

There are three, which are the noblest Creatures in the world, these three bear a wonderful affection one to another. Among Minerals it is man, out of whole *Mume* is made an Animal Stone, in which *Microcosme* is contained. Among Minerals Gold is the noblest, whose fixedness is a sufficient testimony of its noble off-spring and kindred. Among Vegetables there lies hid a Vegetable Stone. Man loves Gold and Wine above all other Creatures, which may be beheld with eyes. Gold loves Man and Wine, because it lets go its noblest part, if spirit of Wine be put to it, being made potable, which gives strength to man, and prolongs his life in health.

Wine bears affection to man also, and to Gold, because it easily unites with the tincture of *Sol*, expels melancholy and sadness, refreshes and rejoices man's heart. He that has these three Stones may boldly say, that he has the Stones of the *Universal*, much of it is talked and written: But what eye has seen it! Not one amongst many hundreds of millons.

These Stones renew men and beasts, cure leprous Metals, cause barren-ness to become fertile, with a new birth, human reason is not able to comprehend it, or conceive of it.

If a rectified *Aqua vitæ* be lighted, then Mercury and the vegetable Sulphur separates, that Sulphur burns bright, being a mere fire, the tender Mercury betakes himself to his wings and flies to his *Chaos*.

He that can shut up and catch this fiery Spirit, he may boast, that he has gotten a great victory in the Chemical battle: for this Vegetable, fiery Sulphur is the only Key to draw the Sulphur from mineral and metalline bodies.

Thus, I close my book, the things contained therein are not grounded on opinion, as most Physicians rely on the Authors, that such and such herbs are cold and moist, dry, and warm, in the first, second, and third degree: because they heard their Authors affirm it, themselves neither saw it, nor made trial of it: Making mere collections from other men's writings, patching up volumes. The things I wrote of, I know by long experimental knowledge, to be true: this my experience I hope will take place, and get the victory as the Amazons did in their prudence.

The eternal heavenly spirit refresh our Souls, that we may walk in heavenly streets, forsaking all false and erroneous by-ways. Amen.

FINIS.

Note: This part ends on page 312 of the original printed book. The next page is numbered 343. I see no evidence that any pages are missing, just a discontinuity in page numbering. -pnw

The First Treatise

Of the

Sulphur, Vitriol, and Magnet

Of the Philosophers.

Section I. Of the Sulphur and Ferment of the Philosophers.

I have written for you, Seeker of that most noble and blessed Art of Chemistry, a small treatise of the Philosophers Stone, divided into XII. Keys, and have expressly enumerated the matter of the Philosophers Sulphur in the first Key, and taught you in the second Key how you ought to distill our water of the *Eagle* and cold *Dragon*, who had his dwelling a long time in Rocky Cliffs, and crept in and out in Subterranean Concaves and Hollow places; pour this Spirit upon purged and fined Gold, let it dissolve and putrefy fourteen days in *Balneo Mariæ* distill it, and pour the Water again on the Gold Calx, and cohobate this until the Gold pass over with the Water, set this again to distill, abstract the Water gently, leave a third part of it in the bottom, then set it into a Cellar, let it Coagulate and Crystallize, wash these crystals with distilled water, amalgama them with Mercury vive, evaporate the Mercury gently, then you have a subtle powder, put it in a glass, lute it, reverberate it for three days and nights, do it gently, thus is the Philosophers Sulphur well prepared for your work, and this is the Purple Mantle, or Philosophic Gold, keep it safely in a glass for your conjunction.

Section II. Of the Philosophers Vitriol.

I have told you plainly how Philosophers Sulphur is made, which *loco masouli*, is to make the King, or Man, now you must have the female, or wife, which is the *Mercury* of Philosophers, or the *materia prima lapidis*, which must be made artificially: For our *Azoth* is not common Vinegar, but is extracted with the common *Azoth*, and there is a Salt made of *materia prima*, or Mercury of Philosophers, which is coagulated in the belly of the Earth. When this matter is brought to light, it is not dear, and is found everywhere, Children play with it: It is ponderous, and has scent of dead body, for two Gilders you may buy this matter for the work: Therefore take this matter, distill, calcine, sublime, reduce it to ashes; for if an Artist want ashes, how can he make a Salt, and he that has not a Metalline Salt, how can he make the Philosophers Mercury?

Therefore, if you have calcined the matter, then extract its Salt, rectify it well, let it shoot into the Vitriol, which must be sweet, without any corrosiveness, or sharpness of Salt. Thus, you get the Philosophers Vitriol, or Philosophic Oil, make further of it a *Mercurial* water, thus you have performed an artificial work: This is called the Philosophers *Azoth*, which purges *Laton*; but is not yet washed, for *Azoth* washes *Laton*, as the Ancient Philosophers have told two or three thousand years ago. For the Philosophic Salt, or *Laton* must with its own humidity, or its own Mercurial water be purged, dissolved, distilled, attract its *Magnet*, and stay with it. And this is the Philosophers Mercury, or *Mercurius duplicatus*, and are two spirits, or a spirit and water of the Salt of Metals. Then this water bears the name of *succus Lunariae, aqua cælestis, acetum Philosophorum, aqua Sulphuris, aqua permanens, aqua benedicta.* Take eight, or ten parts of this water, a:rd one part of your *Ferment*, or Sulphur of *Sol*, set it into the Philosophers *Egg*, lute it well, put it in the *Athanar*, into that vaporous, and yet dry fire, govern it, to the appearance of a black, white, and red color, then you get the Philosophers Stone, and you enjoy this noble, dear, and blessed Medicine and Tincture, and you may work miracles with it.

Section III. Of the Philosophers Magnet.

Hermes the Father of Philosophers had this Art, and was the first that wrote of it, and prepared the Stone out of *Mercury, Sol*, and *Lune*, of the Philosophers: Whom many thousand laborators have initiated, myself also did the like: And I speak really, that the Philosophers Stone is composed of two bodies, the beginning and ending of it must be with Philosophic Mercury.

And this is now *prima materia, alias præda: materia prima* belongs only to God, and is coagulated in the entrails of the Earth, first into Mercury, then into Lead, then into Tin, and Copper, then into Iron, etc. Thus, the coagulated Mercury must by Art be turned into its *prima materia*, or water, that is, Mercurial water. This is a stone and no stone, of which is made a volatile fire, in form of a water, which drowns and dissolves its fixed father, and its volatile mother.

Metalline Salt is an imperfect body, which turns to Philosophic Mercury, that is, into a permanent, or blessed water: And is the Philosophers *Magnet*, which loves its Philosophic *Mars*, sticking unto him, and abides with him. Thus, our *Sol* has a *Magnet* also, which Magnet is the first root and matter of our Stone: If you conceive of and understand my saying, then you are the richest man in the world.

227

Hermes says, three things are required for the work; first a volatile, or Mercurial water, *aqua cœlestis*, then *Leo viridis*, which is the Philosophic *Lune*, thirdly *æs Hermetis*, *Sol*, or *Ferment*.

Lastly note, Philosophers had two ways, a wet one, which I made use of, and a dry one: Herein you must proceed Philosophically, you must well the Philosophers Mercury, and make Mercury with Mercury, adding the Philosophic Salt, Ferment, or Sulphur of Philosophers, and proceed therein, as you heard formerly, then you have the Philosophers *Magnet*, that is, the Philosophers Mercury. Secondly, the Metalline Salt, or Philosophic Salt. Thirdly, *æs Hermetis*, or Philosophic Sulphur.

Thus have I delineated the whole Art, if you do not understand it, then you will get nothing, nor are you predestined thereunto.

Allegorical expressions betwixt the Holy Trinity and the Philosophers Stone.

Dear Christian Lover, and well-wisher to the blessed Art: How graciously and miraculously has the Holy Trinity created the Philosophers Stone. For God the Father is a spirit, and yet makes himself known under the notion of a man, as he speaks, Gen. Chap. 1., let us make man, an image like unto us. *I am*, this expression in his work speak of his mouth, eyes, hands, and feet; so Philosophers Mercury is held a Spiritual body, as Philosophers call him. God the Father begat his only Son JESUS CHRIST, which is God and Man, and is without sin, neither needed he to die: But he laid down his life freely, and rose again, for his brethren and sisters sake, that they might live with him eternally without sin. So is *Sol*, or Gold without defect, and is fixed holds out gloriously all fiery examens, but by reason of its imperfect and sick brethren and sisters, it dies, and rises gloriously, redeeming and tinging them unto eternal life, making them perfect unto good Gold.

The third person in the Trinity is God the Holy Ghost, a comforter sent by our Lord *Christ Jesus* unto his believing Christians, who strengthens and comforts them in Faith, unto eternal life; even so is the spirit of material *Sol*, or of the body of Mercury, when they come together, then is he called the *double Mercury*, these are two Spirits, God the Father, and God the Holy Ghost: But God the Son, a glorified Man, is even as our glorified and fixed *Sol*, or Philosophers Stone; since *Lapis* is called *Trinus*, namely, out of two waters, or Spirit of Mineral, and of Vegetable, and of the Animal of Sulphur of *Sol*: These are the Two and Three, and yet but one: if you understand it not, then you are not like to hit any. Thus by way of a similitude I delineated unto you sufficiently the *Universal*. Pray to God for a blessing, for without him you are not like to prosper at all.

The Second Treatise

Of Vulgar

Sulphur, Vitriol, and Magnet

Section I. Of Sulphurs.

Chapter I. Of Sulphur of Saturn.

1. There is extracted from calcined Saturn with distilled Vinegar a Crystalline Salt, which is distilled with Spirit of Wine unto a red Oil. This Oil cures Melancholy, fiery Pox, old Ulcers, and many other infirmities besides.
2. This Oil coagulates and fixes Mercury, being first precipitated. with Oil of Vitriol, for all Powders and Medicaments, which are to make *Sol*, and *Lune*, must be made fix, holding in all fiery trials.
3. Out of this Oil is made a glorious Tincture, if you take three parts of Mercury of Mars, and one part of this red Oil of Saturn, joining, coagulating, and fixing them: This work may be accomplished in a month's time, or somewhat longer. This Tincture may be augmented with Mercury of *Mars*, *usque in infinitum*; projecting one part of it upon three parts of *Sol*, to make thereby an *ingresse* for the Tincture; one part of this Tincture transmutes thirty parts (if so be it be well prepared) of *Mercury* and of *Lune* into good *Sol*. Remember your Creator, and be mindful of the poor, then the Lord will be mindful of you also.

Chapter II. Of Sulphur of Jupiter.

1. There is made of *Jupiter*, a Salt like that which is made of Saturn, from thence is extracted and distilled a red sweet Oil: this Oil tinges Saturn, being first calcined with Sal armoniack into *Sol*.
2. The *Limature* of *Jupiter* being calcined with *Calx vive* for a day, and the *Calx* being washed from it, then you have a fixed powder, if you can reduce it

again into a fluidness, and separate it with Saturn, then you may get enough of Gold and Silver.

3. There is a calcination made of Tin and Lead with common Salt, but better is it, if made with Salt of the *caput mort* of Vitriol and Peter[19], the Oil of Vitriol being added unto Calx of *Jupiter* and *Saturn*, and made one mass of it, lute it well , let it stand for eight days and nights in warm sand, and then drive it: one Centuer of Lead affords in this manner six Mark and a half of fine *Lune*: One mark of such *Lune* yields two drachmes of *Sol*.

This has been my first piece to make *Lune* and *Sol* withal. Note, these six Mark and a half of *Lune* afford thirteen drachmes of *Sol*, this *Sol*, and *Lune* amounts to 208 Cildors or 20 pound and 16 shillings.

Chapter III. Of Sulphurs of Mars and Venus.

1. Take some pounds of Verdigreece, extract its Tincture with distilled Vinegar, let it shoot, then you have a glorious Vitriol, out of which you may distill *per retort* a red Oil. This Oil dissolves *Mars* turning into a Vitriol, which is reduced in a long time in a great fire into a red Oil, then you have together Sulphur of *Mars* and *Venus*, add somewhat of Sulphur of *Sol*, coagulate and fix it, then you have a Medicine, which meliorates Men and Metals.
2. *Lune* is graduated with this Oil, and a good part of the Kings Crown is gotten.
3. Two equal parts of laminated *Sol*, and *Lune* melted together, putrefied in this Oil for a day and night, turns them into good Gold. In this Oil you will find many strange effects and virtues.

Laus Deo

Chapter IV. Of Sulphur of Sol.

1. I have formerly told how Gold is made Spiritual unto the Purple Mantle. Now if you are about to make *aurum potable*, that you may dissolve with the Oil of Vitriol that spiritual Gold, extract, and draw it over with Spirit of

---

[19] Saltpeter?

Wine: This is a Medicine, which cures many difficult diseases, and is wonderful in efficacy.

2. This *Solar Sulphur* tinges prepared *Calx* of *Lune* into good Gold, but you heard in my former expressions, that the King has only an honorable Garment, and must raise his Rents and Subsidies from his Subjects, must be bathed in his flourishing blood and sweat, must be destroyed. and gloriously renewed, then is he able to make his poor brethren and sisters to be Kings also, and legitimate their bastards.

Antimony is a bastard of Saturn, how much of *Regulus* it has, so much is turned into *Sol*, its due Dose of Tincture being first added thereunto. *Marcasite* a bastard of *Jupiter*, is turned to *Sol*, also by means of a tincture. Vitriol has in it a Metalline Mercury, a bastard of *Venus*, as much as it has of it, so much is tinged into *Sol*.

3. If you add the Solar Sulphur unto Sulphur of Vitriol, *Venus* and *Mars*, and then fix them artificially, then you have a tincture for Men and Metals, expelling all manner of diseases: And this fixed powder tinges *Particulariter Lune* into *Sol*.

Laus Deo.

Chapter V. Of Sulphur of Mercury.

Mercury is the Original of all Metals, and is a Spiritual body, and a fugitive servant, when it is cast into the fire, then it flies into his *Chaos*. But he that can catch him, he gets then the Sulphur of Mercury, or Water of Sulphur, or *aqua benedicta*, the Key of the Art, which opens Metals Philosophically. The Philosophers Mercury, and not the vulgar, being reduced unto water, dissolves the Philosophic Salt together with the Purple Mantle, by putrefaction and distillation, for it is *Mercurius duplicatus*.

Chapter VI. Of Sulphur of Lune.

1. This *Lune* is made spiritual by means of our water, of the Second Key, and may easily be made into potable Silver, whereby many diseases are cured.

2. Feed three parts of this Spiritual *Lune* with equal parts of Virgins Milk, and bring two of them unto fixation, then you have an augment of *Lune*, which breeds monthly young ones, these are taken forth, and their places are supplied with Mercury vive, *&c*. This powder is reduced with Borax, then you have an *augmentum perpetuum*.

Chapter VII. Of Sulphur of Antimony.

1. There is made out of *Vitrum Antimonii* with distilled Vinegar a sweet extraction, let the *acetum* be separated from it: on the remainder pour spirit of Wine, which must be extracted, and the pure from the impure separated. This sweet extraction draw over, let the spirit of wine by cohobation be often drawn from it, and that powder may be reduced to a glorious Oil of Antimony.

This Oil cures all manner of diseases, being ministered in a convenient vehicle.

2. Further, take one part of this Oil, and two parts of the Mercurial water, in which is dissolved a fourth part of the purple Mantle, then dissolve, conjoin, lute Hermetically, coagulate, and fix. This Tincture tinges *Lune* and *Mercury* into *Sol*. This is that pure Sulphur of Antimony, which is made of the Vitrum of Antimony *per se*, without any addition of Saltpeter, Salt or Borax.

Chapter VIII. Of Sulphur of Vitriol.

1. There is made of Vitriol a *lixivium* with ashes of Beech-wood, and a Sulphur is drawn from this Vitriol, and is precipitated with Salt of Tartar; then the red Oil or Sulphur is extracted with *Jupiter* Oil,[20] putrefy the same with spirit of Wine, abstract the spirit of Wine from it. This glorious Oil of Sulphur is good against many diseases, it is to be used against Consumption, Dropsie, Plague, Scabbiness, and such like.
2. Vitriol is sublimed with Sal armoniac also, and better than if done with a *lixivium*, because the body of Vitriol is better opened and dissolved. This sublimate is dissolved into an Oil, whereby crude Mercury can be coagulated and fixed, of the which I shall write more anon, when I treat of Vitriol.

[20] Juniper Oil?

Chapter IX. Of common Sulphur.

1. There is a Liver made of yellow Sulphur with Linseed Oil, then decoct and putrefy it in a *lixivium* made of Sal Alcaly, and then distill it; pour this water on Tyles, which newly came out of the Furnace, imbibe them therewith, distill it *per retortam*, you have a yellow water of it like an *aquafort*, which tinges *Lune*. Take one part of this water, and one part of *Lunar calx*, let it stand three days and nights in warm sand, the fourth part of it turns into Gold, be reduced, separated, and purged with SATURN.

2. Further, Sulphur with the *anima* of *Saturn*, being often driven over and fixed, may then safely be used inwardly for a Medicine: But projected on *Lune*, in the flux it affords good Gold a fourth part.

3. Of the Gray powder and *Calx vive* equal parts, one pound, a fourth part of Salmiac ground therewith, and driven over *per retortam*, affords a glorious red Oil, which is of a fixing and graduating quality.

4. Lastly, I tell you, take of this Oil of Sulphur, of *Venus*, and of *Mars*, add thereunto the Oil of the Sulphur of Antimony, bind these together with the Oil, or Water of Mercury, fix it, then you have a Medicine for Men and Metals, *viz.* to tinge Mercury and *Lune* into *Sol*.

Section II. Of Vitriols.

Chapter I. Of Vitriol of Sol, and of Lune.

In the first place you must have our water of the cold *Earth Salt*, and of the *Eagle*, wherewith Gold and Silver are made spiritual, and coagulated into crystals, or a Metalline Vitriol; out of which is distilled withsSpirit of wine, after its proper manner, an Oil of Sulphur, to be used after the manner of Metal.

Chapter II. Of Vitriol of Saturn, and of Jupiter.

Calcine Saturn, or Jupiter, extract its *animæ* with distilled Vinegar, let it putrefy fourteen days, let the Vitriol shoot. This must be driven over with Spirit of Wine, it affords a sweet Oil, and. it is the Sulphur of *Saturn* and *Jupiter*. This Oil coagulates Mercury, and being first precipitated with Oil of Vitriol , it fixes it.

Chapter III. Of Vitriol of Mars.

Take the filings of Mars and Sulphur equal parts, calcine them in a Brick-kiln to a purple color, pour on it distilled Water, or Vinegar, it extracts a green color, abstract two parts of that water, let it shoot: thus you have a noble Vitriol, distill from it, a red Oi , take an ounce of it, add to it Mercurial water, in which is dissolved *Sol*, fix this Tincture, Lead, Silver, and Tin, are transmuted thereby into pure Gold. Return thanks to God the Creator of Minerals, Metals, and all other Creatures!

Chapter IV. Of Vitriol of Venus.

1. I have taught you already to extract the transparent Vitriol out of *Venus*, and to distill its red Oil. This Oil dissolves *Mars*, into Vitriol, and being once more distilled *per retortam* forcibly, then you have an excellent Tinging Oil, or Salt of *Mars*. This is the Kings Excise man, who brings in his Rents, and enriches him. This Oil dissolves the spiritual purple Gold, and draws it over the Helmet. Now you have fermented the *Solar Sulphur* with its own Sulphur, which Philosophers before me have not done, but fermented calcined Gold, or the calx of Gold in *Mercurius Duplicates*, and attained unto the end they wished for, as well as I. But according as men do work, so is the operation of

their Tincture, transmuting more, or less, according to the efficacy of the Tincture.

2. Out of the Oil of this *Martial* Salt, is Mercury of Antimony precipitated, and added to the sweet Oil of Vitriol, is fixed, this medicine next unto the Philosophers Stone is the best, and highest *Universal* upon man's body, and tinges *Lune*, *Saturn*, and *Jupiter* into good *Sol*, holding in the examen very well.

3. There is made also a mass out of Honey, Salt, and Vinegar, and lamens of *Venus*, which are stratified and calcined. This *calcinate* of its own accord, turns to a Verdigreece, which must be extracted, crystallized, and distilled to a red Oil; which is used as you heard above.

Chapter V. Of Vitriol of Mercury.

1. Vitriol of Mercury is easily made, with *aquafort* made of Saltpeter and Allome being dissolved therein: Crystals do shoot like unto Vitriol; these being washed distill them into a sweet Oil, with spirit of Wine first certified[21] with Salt of Tartar, this is an excellent Medicine against the *French* disease, old Ulcers, Cholick, Windy-ruptures, Gouts, expelling many other diseases out of man's body.

2. This Oil is joined also with *Martial* Tinctures. For Mercury is the bond of other Metals, and may be well used *Particulariter*. The chiefest color of Mercury is red, that is after it is precipitated as you find in my other writings.

Chapter VI. Of Common Vitriol.

1. Take a good *Hungarian* Vitriol, dissolve it in distilled water, coagulate it, let it shoot into Crystals: iterate it five times, then is it well purged, and the Salts, Allums, and Niter are separated from it. Distill this purged Vitriol with spirit of Wine into a red Oil, ferment with Spiritual Sol, add to it a due Dose of quick Mercury of Antimony, coagulate, and fix, then you have a Tincture for men, and it tinges *Lune* also into *Sol*.

> *Visitando Interiora Terræ*
> *Rectificandoque,*
> *Invenietis occultum Lapidem,*
> *Veram Medicinam.*

---

[21] Clarified?

2. Vitriol is calcined also to a red color in a close Vessel, on which pour distilled Vinegar, and set it in putrefaction for three months, there is found in a strong distillation, a quick Mercury, which you are to keep safe. Wonders may be effected therewith, in *Particulars* and *Universals*. Take three parts of this Mercury, and one part of *Sol*, join these, which being fixed affords a *Solar augmentum*. Make your supplies with its Mercury. *Laus Deo*.

3. This calcined red Vitriol is sublimed also with Sal armoniac, this sublimate is dissolved into an Oil. This fixes Cinebar, whereof may be had *Lune* and *Sol*.

4. There is made a fixed water also, Sal armoniac and Allom being added thereunto. This water being poured upon *Sulphur* of *Saturn*, which before was precipitated into a red powder, imbibe, and coagulate, and let an *ingresse* be made with *Sol*, then you have a tincture, whereby crude Antimony is transmuted into good *Lune*, which may be transmuted into *Sol*.

5. Lastly, I tell you, if you extract the Salt out of Vitriol, and rectify it well, then you have a work which is short, and tinges *Lune* into *Sol*, this Metalline Salt coagulates vulgar *Mercury*, and being transmuted into *Lune*, it may be graduated higher by Antimony.

Thus you have my operation and experiments, which may be very profitable unto you. Make a good Christian use of it, help the poor, cure the diseased, then God will bless you. AMEN.

Sulphur is Vitriol
Antimony is Mercury.

Section III. Of vulgar Magnet.

1. Magnet contains that, which common *Mars* has. Common Iron may easily be wrought, I need not to make many words of it. Magnet has an attractive quality to draw Iron.
2. There is made an Oil of *Magnet* and *Mars*, which is very effectual in deep wounds.
3. With Magnet and Antimony is made *Lune* fix, which with the Oil of *Mars* and *Venus* is graduated, and made Gold: it may be performed also with Antimony and Mars.

Thus I finished my course, and found many things in my working. My fellow brethren turned Alchemists, all had the Philosophers Stone. I was the beginner, took great pains, before I attained to anything, if you read my writings diligently, you will find in the Parable of the XII. Keys, the *prima materia*, or Philosophers Mercury, together with the Philosophic Salt: the Philosophic Sulphur, or ferment, I delineated expressly.

Now I close, and commit you to God, and desire you to remember me.

Medium Tenuere Beati.

FINIS.

Jod. V.R.

A Process upon the Philosophic work of Vitriol.

Having gotten this Process in the foresaid year, and afterward, as you shall hear, with mine own hands elaborated and wrought the same, no man over-looking me, I was heartily rejoiced, even as if I had been new born, and returned hearty thanks to God: Its practick at the first I have not plainly described, because I had erred in the composing of it, and was fain to begin the work anew. I having missed in my work, I began in the year 1605. Because the matter of the Earth, and the Spirit of Mercury was not sufficiently purged, therefore the earth could not perfectly be united at the composition with the water, I let that quite alone, and began a new Process at the end of the year 1605 in the City of *Strasburg*, used more diligence and exactness, then my work (God be praised) prospered better, for the which I am still thankful to God for it. In the name of the Holy Trinity, the 19th. of *October*, Anno 1605. I took ten pounds of Vitriol, dissolved it in distilled Rain-water, being warmed, let it stand for a day and a night, at that time many *feces* were settled, I filtered the matter, evaporated it gently, *ad cuticulum usque*, I set it on a cool place to crystallize, this onshot Vitriol I exiccated, dissolved it again in distilled Rain-water, let it shoot again, which work I iterated so long till the Vitriol got a cœlestial green color, having no more any *feces* about it, and lost all its corrosiveness, and was of a very pleasant taste.

This highly putrefied Vitriol, thus crude, and not calcined, I put into a coated Retort, distilled it in open fire, drove it over in twelve hours space by an exact government of fire in a white fume, when no more of these fumes came, and the red corrosive Oil began to come, then I let the fire go out, the next morning, all being cold, I took off the receiver, poured the gift in the receiver into a body, and some of the lute being fallen into, I filtered it, and had a fair menstrual water, which had some phlegm because I took that Vitriol uncalcined, which I abstracted in a *Balneo*, not leaving one drop of water in it.

I found my *Chaos* in the bottom of a dark redness, very ponderous, which I poured into a Viol, sealed it *Hermeticè*, set it on a three-foot into a wooden globe into a vaporous bath made of water, where I left it so long; till all was dissolved, after some weeks it separated into two parts, into a bright transparent water, and into an earth, which settled to the bottom of the glass, in form of a thick black corrosive, like pitch. I separated the white spirit from it, and the fluid black matter I set it again to be dissolved, the white Spirit which was dissolved of it, I separated again, this work I reiterated leaving nothing in the bottom, save a dry red earth. After that, I purged my white Spirit *per distillationem* very exactly, it was as pure as the tear that falls from the eye, the remaining earth I exiccated under a muffle, it was as porous, and as dry as dust, on this I poured again my white spirit, set it in a digestion, this spirit extracted the Sulphur, or Philosophic Gold, and was tinged of a red yellow, I canted it off from the matter, and in a body I abstracted the spirit from the Sulphur, that Sulphur stayed behind in form of an Oil, very fiery, nothing like unto its heat, as red as a Ruby: this abstracted white Spirit I poured on the earth again, extracted further in Sulphur, and put it to the former. After this that *Corpus terræ* looked of a paler colour, which I calcined for some hours under a Muffle, put it into a body, on it I poured my white spirit, extracted its pure white fixed Salt, the remaining earth was very porous, good for nothing, which I flung away, thus these three principals were fully and perfectly separated.

After all this I took my astral clarified Salt, which weighed half an ounce, after the weight of *Strasburg*, and of the white spirit, which weighed four ounces, of Mercury one ounce, and a quarter of an ounce, these I divided into two parts, whose quantity was half an ounce and one dram, I put this Salt to one part of the water in the Viol, and nipped it, set it in digestion, there I saw perfectly how the Salt dissolved itself again in this spirit, therefore I poured to it the other part, which was half an ounce and one dram, no sooner this was put to it, then presently the body together with the spirit turned as black as a Coal, ascended to the end of the glass: and having no room to go further, it moved to and fro, sometimes it settled to the bottom, by and by it rose to the middle, then it rose higher, thus it moved from the fourth of *July*, to the seventh of *August*, namely thirty four days, which wonderful work I beheld with admiration; at last these being united , and turned to a black powder, staying on the bottom, and was dry, seeing that it was so, I increased my fire in one degree, took it out of the wet, and set it in ashes, after ten days the matter on the bottom began to look somewhat white, at which I rejoiced heartily, this degree of fire I continued, till the matter above and below, became as white as the glittering Snow. But it was not yet fix, making trial of it, set it in again, increased my fire one degree higher, then the matter began to ascend and descend, moved on high, stayed in the middle of the glass, not touching the bottom of it, this lasted thirty-eight days and nights, I beheld then as well, as formerly at the thirty days a variety of colors, which I am not able to express. At last this powder fell to the bottom, became fix, made projection with it, putting one grain of it to one, and a quarter of an ounce of Mercury, transmuting the sane into very good *Lune*. Now it was time to restore unto this white tincture her true *anima*, and imbibe it, to bring it from its whiteness unto redness, and to its perfect virtue.

Thereupon I took the third principle, namely the *anima*, which hitherto I had reserved, in quantity it was one ounce, a quarter of an ounce, and one dram, poured to it my reserved spirit of Mercury, whose quantity was one ounce and a quarter of an ounce, drew it over several times *per alembicum*, so that they in the end united together; those I divided into seven equal parts: one part I poured on my clarified earth, or tincture, which greedily embraced its *anima* together with its spirit, and turned to a ruddiness in twelve days and nights, but had no tinging quality as yet, saving Mercury vive and Saturn, it transmuted into Lune, which Lune at the separating yielded three Grains of Gold. I proceeded further with my imbibition, and carried all the seven parts of *anima* into: At the fourth inbibitation one part of my work tinged ten parts of Copper into Gold, at the fifth inbibitation one part tinged a hundred parts, at the sixth it tinged a thousand parts, at the seventh it tinged ten thousand parts: Thus, God be praised, my work ended successfully, with great joy of my heart: At this time I got of the true *Medicine* four ounces half an ounce, and one dram. The two last in the ponderosity were almost equal unto the first, out of this my work I paid for land and ground, to that nobfe Gentleman O.V.D. 48,000 Gilders, *Actum* 1607. These things I set down for a *memorandum*, that I should not forget any of the manuals, and of other things necessary for the work. God be praised for evermore. AMEN.

Coagulation of Mercury Vive into a Lunar Fixation

An exact work, how Mercury Vive is coagulated and brought unto a LUNAR fixation, which LUNA holds SOL also in the trial.

Take of *Mercury vive* two ounces, of pulverized common Sulphur six ounces, grind these in a wooden dish with a wooden pestle , set it on a coal-fire in a melting pot, stirring it about continually, let all the Sulphur evaporate: then take forth the Mercury, grind an equal quantity of Sulphur with it, proceed with its heating as formerly: iterate this work five times, then sublime this Mercury *per gradus ignis*. Take out this *sublimate*, break it in pieces of the bigness of a small Nut or Bean, imbibe them in the white of eggs, then take a cementing pot, put ashes into it, in the midst of it set an iron box, stratify into it this *sublimate* with refined Silver, fill up the box, then lute an iron lid to it, put ashes on the lid, lute an earthen lid upon that, set this pot into a sand Capel, let your first fire be gentle for twelve hours, then increase your fire for twelve hours more, at last make a forcible fire for twenty-four hours, then break open the Pot, you will find a black grey matter, carry it on Lead, of four ounces you will get three ounces of fix Silver: separate this fine Silver in *aquafort*, you will find a good deal of black Gold Calx, reserve the Silver Calx apart, you may stratify with it another time. Thus far I went in my experience.

The Fifth and Last Part

Of the Last

TESTAMENT

Of Friar

Basilius Valentinus

The Transcendent Medicine

Treating of the Transcendent, and most precious and wonderful Medicine, which the great Creator has put into Metalline and Mineral Salts, for the benefit of Man: To keep him in perfect health continually.

Before I begin to speak of the Salts of Metals and Minerals, and declare their volumes, and other precious and noble growths underground, in the first place I will prefix the preparation of *aurum potable*, because therein lies the *Corona* of Medicinals, *Universaliter*, and merits the first place, because Salts of other Metals and Minerals in their innate virtues are for *particulars* only, and are ordained for to preserve man in health; and there is just cause to begin with the making of *aurum potable* without sophistication, and will speak of the distinction of it, that it may be judged infallibly to be the true one.

This being my last part, and my intention is to make a perfect relation of *aurum potable*, for the benefit of good and understanding men, to whom God after my death will bestow this my book, which upon tedious and laborious experience I wrote, wherein I speak not by hearsay, but the things I do write of, I know experimentally to be true.

Therefore, if God does bless you with a true knowledge hereof, that you would keep this *secret* in silence and privacy, least you turn Gods blessings into a curse: Because the preparation of this, and of the Stone is one, both have their original, and first generation and birth from the true seed, and Astrologic *primum mobile*, called the spirit of Mercury, of which formerly I have written more largely. For I speak the highest truth unto you, that neither the *Universal* nor *Particular Tincture*, neither *aurum potable*, nor other *Universal Medicine*, without this heavenly and spiritual essence, which has its original from the starry heaven, taking and receiving the same, from thence

242

may be had and prepared, therefore be silent till death, at your departing lay down your talent; as I have done; for if I had not informed you faithfully, you would know but little of that mystery, and continue still with the vulgar in folly, blindness, and madness, and you would have sent a Recipe into the greasy and salvy shops of Apothecaries, but whither would your Soul have gone after your departure? Into *Galens* lap, to the utmost depth of darkness, where the Devils have their dwelling places, even thither, both your soul and body would have been sent, in case you should have divulged any of these secrets.

To turn to my intended business, I will in the first place inform you, what is that true and highest *aurum potable*, and *Universal Medicine*, after this in order there follows another *aurum potable* made of the fixed red *Sulphur*, or Soul of the corporeal Gold, most highly purged, and is prepared with the conjunction of the *Universal Spirit of Mercury*. After this there follows another *Particular Medicine*, which is half an *aurum potable*, showing its efficacy and power in many trials. Then I will add thereunto a description of *aurum potable*, because it traces the steps of Gold, and it shows wonderfully its great energy and virtues.

The highest and chiefest *aurum potable*, which the Lord God has laid into nature, is the excocted, prepared, and fixed substance of our stone, before it is fermented. A higher, greater, and more excellent *Universal Medicine*, and *aurum potable* cannot be found, nor had in the circumference of the whole World; for it is a heavenly Balsame, because its first principles, and original comes from heaven, made formal in earth, or underground, and is afterwards, being exactly prepared, brought into a *plusquam perfection*, of which first principle and Nativity of this heavenly substance I have already written sufficiently, and count it needless to be repeated here.

Now as this excocted and perfect substance is the highest, chiefest, and greatest *Universal Medicine* unto man, even so on **the** other side the same matter after its *fermentation*, is a Tincture also, and the chiefest, greatest, and most powerful *Universal Medicine* upon all metals whatsoever, and thereby may be transmuted into their highest melioration and health, namely into the purest Gold. This is the first, chiefest, and greatest *aurum potable* and *Universal Medicine* of the whole World, of which alone great volumes could be written; whose preparation was set down circumstantially in the third part, needless to be repeated here again; At the present I will speak of the true and full *process*, how a true *aurum potable* is to be had, and prepared from Gold, which in the best manner is most exactly putrefied. Take the extracted Soul of Gold, draw forth with the sweet spirit of common Salt, as I informed you about the *Particular* of Gold, where the body of Gold appeared very white, abstract the spirit of Salt from it, edulcorate the *anima* of *Sol* ten or twelve times; at last let it be purely exsiccated, weigh it, pour on it four times as much of *spirit* of *Mercury*, lute it well, set it in the vaporous Bath, putrefy it gently, let the *anima* of *Sol* be quite dissolved, and be turned into a water, or its *prima materia*, both will turn into a blood-red liquor, fair and transparent, no Ruby on the earth is comparable unto it.

But thus much you must note, when the *anima* of *Sol* begins to be dissolved, and brought into its *prima materia* that at the first, on the side round the glass, where the matter lies, there be seen a green circle, on it a blue, then yellow, afterward all the colors of a Rainbow, join, and make appearance, which do last but a little while. The *anima* of *Sol*, being wholly dissolved into the *Mercurial Spirit*, and nothing is seen in the bottom, then pour to it twice as much of the best rectified Spirit of Wine, brought to its highest degree, the glass must be luted exactly, digest, and putrefy gently for twelve, or fifteen days, together, then abstract *per alembicum*, that matter comes over in a blood-red transparent color: this abstracting must be iterated, nothing must be left in the bottom, which is corporeal, then you have the true *aurum potable*, which can never be reduced into a body.

But note, the Gold before destruction and extraction of its Soul must be purged in the highest degree. There is made another *aurum potable*, and artificially prepared, which though it cannot be said, or set down in writing to be the full and true potable Gold, yet it is more than half an *aurum potable* counted, because it is transcending effectual in many diseases, in which nature might have stood in great doubts. This half *aurum potable* is made in a twofold manner, where the latter is better and more effectual than the former, and asks more pains and time than the former.

Take this extracted Soul of Gold, drawn forth with the sweet spirit of common Salt, edulcorate it most purely and exactly, at last exsiccate it, put it in a spacious Viol, or body of glass, pour on it red Oil of Vitriol, which was dephlegmed, and rectified *per retortam*, that it be transparent, clear, and white, and you may see, that it seizes on the Gold and dissolves it, and is tinged deeply red.

Put so much of this Oil to it, that in it may be dissolved Sulphur, or the Soul of Gold, let it putrefy in *Balneo Mariæ*, put a reasonable fire to it, that you may see that the Soul of Gold is quite dissolved in the Oil of Vitriol: The *feces*, when it has settled, must be separated from it, then put twice as much of the best rectified spirit of Wine to it, which rectification you shall be informed of in this part, seal the glass, let no spirits of the Wine evaporate, set it again in putrefaction in the Balneo, let it be there for a month, then the sharpness of Vitriol is mitigated by the spirit of Wine, and loses its acidity and sharpness, both together make an excellent Medicine, drive both over, let nothing stay behind in the bottom, then you get more than half an *aurum potable*, in form and color of a deep yellow liquor. Note, that some Metals in this manner may be proceeded withal, first a Vitriol may be gotten out of the Metal, then a spirit is further driven from it, and joined in this manner with the Soul, dissolved, and further digested with spirit of Wine, all must enter together into a Medicine, as I told of formerly, which have their special efficacy.

The second way to prepare this half *aurum potable*, which though it be but half an *aurum potable*, yet in virtue and efficacy is far preferred before the other now spoken of, and is done as follows:

Take the extracted *Solar* Soul spoken of above, put it into a Viol, pour on it the extracted Philosophic Sulphur, which is the second *principle*, which is drawn with spirit of Mercury from the Philosophic earth, and Mercury, or the spirit of Mercury, unto an Oleity, which now is Sulphur again, and must be abstracted gently *per molun distillations*.

Of this Philosophic Sulphur pour on it as much, that the *Solar* Soul may be dissolved, let it stand in a gentle Bath, let the dissolution be made, then pour more of the best spirit of Wine to it, digest gently, draw these over, let nothing stay behind in the bottom, then you have a Medicine, which does not want above two Grains of the right and true *aurum potable*.

These are the chiefest ways to make the corporeal *aurum potable*, this I close and proceed further with a short, but true process, how the Silver, which is the next to Gold, concerning perfection, is made potable also: this process must be done in the following manner.

Take the sky-colored Sulphur, or spirit of *Lune*, which was extracted with distilled Vinegar, as I informed you in the *Particular* of *Lune*, edulcorate it, rectify it with spirit of Wine, exsiccate it, put it in a Viol, pour to it three times as much of spirit of Mercury, which is prepared from the white spirit of Vitriol, as I faithfully taught you in that place, lute the glass firmly, set it in putrefaction in the vaporous Bath, let all be dissolved, and nothing more seen in the bottom, then put to it an equal quantity of this spirit of Wine, set it in digestion for half a month, drive all over, let nothing stay behind, then you have the true potable *Lune*, which in its efficacy is admirable, and does wonders when it is used.

A description of the Fiery Tartar.

Distill of good Wine a spirit of Wine, rectify it with white calcined Tartar, let all come over, put that which is distilled over into a Viol, put four ounces of well sublimed Sal armoniac to one quart of spirit of Wine, set a Helmet upon, set a great Receiver into cold water, drive the volatile Spirits into, gently in *Balneo Mariæ* leave but a little quantity of it behind. Note the Alembic must always be cooled with wet cloths, then the spirits will be dissolved, and turn to liquor. Thus is prepared this hot spirit of Wine.

Of the Salt of Tartar.

First you must note, that the Philosophers Tartar is not the vulgar Tartar, wherewith the Lock is opened, but it is a Salt, which comes from the root, and is the only mystical Key for all Metals, and is prepared thus: make a sharp *lixivium* of the ashes of *Sarments,*[22] or of twigs of the Vine, boil away all its moisture, there stays behind a ruddy matter, which must be reverberated for three hours in a flaming fire, stirring it still, let it come to a whiteness, which white matter must be dissolved in distilled Rain-water, let the *feces* of it settle, filter, and coagulate them in a glass, that the matter in it be dry, which dry matter is the Salt of Tartar, from which the true spirit is driven.

Note, as I told now of the virtue and qualities of precious stones, so there are found also many despicable, and ignoble stones, which are of great virtues, and experimentally are known to be of rare qualities, which ignorant, and inexpert men will hardly give credit unto, neither can they conceive of it in their dull reason and understanding: the same I will demonstrate with the example of *Calx vive*, which in men's judgement is held of no great value, and lies contemptibly in obscurity, however there is a mighty virtue and efficacy in it, which appears, if application be made of it to the most heaviest diseases, seeing its triumphant and transcendent efficacy is almost unknown for the generality, therefore for the good of such, which are inquisitive into natural and supernatural mysteries, and to whom I disclose these mysteries in this my book, I will for a farewell discover also this mystery concerning the *Calx vive*, and will show in the first, how its spirit is driven from it, which work indeed requires an expert Artist, who is well informed aforehand of its preparation.

Take unslaked Lime, as much as you will, beat, and grind it on a well-dried-stone, to an impalpable powder, put on it so much of spirit of Wine, as the pulverized Calx is able to drink, there must not stand any of that spirit upon it, apply a Helmet to it, lute it well, and a put a receiver before it, abstract the spirit gently from it in *Balneo*, this abstracting must be iterated eight, or ten times: this spirit of Wine strengthens the spirit of Calx mightily, and is made more fiery hot. Take the remaining Calx out of the body, grind it very small, put to it a tenth part of Salt of Tartar, which is pure, not containing any *feces*.

As much as this matter weighs together, add as much of the additional of Salt of Tartar thereunto, namely the remaining matter, from which was extracted the Salt of Tartar, and it must be well exsiccated, all this must be mingled together, and put in a well-coated Retort, three parts of the Retort must be empty, take a great Receiver, or body to it, very strongly. Note, the body into which the Retort's Nose is put, must

---

[22] Garments?

have a Pipe of a fingers breadth, unto which may be applied another body, and a quantity of spirit of Wine in it: then give a gentle fire to it, at first there comes some of the phlegm, which falls into the first applied body: the phlegm being all come over, then increase the fire, there comes a white spirit to the upper part of the body, like unto the white spirit of Vitriol, which does not fall among the phlegm, but slides through the pipe into the other body, drawing itself into the spirit of Wine, embracing the same, as one fire does join with the other.

Note, if the spirit of Calx be not prepared first by the spirit of Wine, and drawn off and on, as I told, then he does not so, but falls among the phlegm where he is quenched, losing all its efficacy. Thus difficult a matter is it, to search nature thoroughly, reserving many things unto herself. This spirit being fully entered into the spirit of Wine, then take off the body, put away the phlegm, but keep carefully the spirit of Wine, and spirit of Calx: and note, both these spirits are hardly separated, because they embrace closely one another: and being distilled, they come over jointly.

Therefore, take these mixed and united spirits, put them into a Jar-glass, kindle it, the spirit of Wine burns away, the spirit of Calx stays in the glass, keep it carefully. This is a great *Arcanum*, few of other spirits go beyond its efficacy, if you know how to make good use of it. Its qualities nay hardly be set down in any way of abridgement.

This Spirit dissolves *Oculi Canororum*, the hardest Crystals: These three being driven over together, and often iterated in that distilling, three drops of that liquor being ministered in warm Wine, break, and dissolve any Gravel and Stone in man's body, expelling their very roots, not putting the patients to any pain.

The spirit of Calx at the beginning looks bluish, being gently rectified, looks white, transparent, and clear, leaving few *feces* behind. This Spirit dissolves the most fixed Jewels, and precious stones. On the other side, he fixes all volatile spirits with his transcendent heat.

This Spirit conquers all manner of Podagrical Symptoms, be they never so nodose and tartarous, dissolves and expels them radically.

To the omnipotent Trine God, Father, Son, and Holy Ghost, be returned hearty thanks for all his benefits, which he has bestowed on man, and discovered those secrets, I wrought on in his name. To him be eternal praises. AMEN

ALL THAT HATH BREATH, PRAISE THE LORD.
ALLELUJAH!

End of the fifth part.

His Treatise Concerning the Microcosme

Or

The little World, which is Man's body.

**What it contains, and of what it is
Composed, what it does comprehend,
And its end and issue.**

**A thing most necessary, and meet for the
Knowledge of such, that love, and
Embrace wisdom.**

Those that seek Art, and have a desire to attain to wisdom, are to note, that the Highest, upon my continued prayers has granted unto me a Clergy-man, to make known the many and great mysteries of Nature: among which man's body is one to be considered, how that is governed in imitation unto *Microcosme*. For it is meet that the lesser should imitate the greater, and the smallest and meanest ought to be governed by the greatest and most potent.

Microcosme, or the great World contains three things, as the most principle, the rest, which came from these, are merely accidental. In the first place is to be considered the matter and form of this World, which matter is made formal out of a non-shape, or a nothing, and the great Creator presently prescribed an order for this matter, what government it should keep, as soon as it came to a life, or motion. This matter and form is water and earth. For at the Creation, by a separation of the water from the earth, there was finished the matter and form, as two things belonging one to another, from these all Animals and Vegetables have their beginning, and other two things, as air and fire, which belong, one to another, have wrought life therein. The matter and form is earth, the Salt in that earth, the body: Even so is it with man's body, which is *Microcosmus*. The matter was not perfect without the form, these joining into one, by God's ordinance, the form being become quick, came then to a perfection: the matter and form got life by motion, air was the first cause of that motion, and perfect maturity was caused by a convenient heat, moveably enclosed in the air, thus the earth was brought to a fertility by the air, it was opened, and became porous by motion for generation. The earth being impregnated, made her seed apparent by her aquosity, then air and heat in the nether and upper Region of the Astrals caused that a birth was brought forth, the blossoms were produced, and the appointed fruit was ripened by concoction of heat.

Calcidity is a Sulphureous hot spirit, which like a Medicament exsiccates the superfluous gross aquosity and phlegmatic matter, which in the generation at the beginning abounds too much in the earth, before the air could have a fellow dominion at the joining with it, carrying the same along in the superfluity of her birth.

The second principle part of the *Microcosme* is *inability*, for the matter in itself was without life, which by heat was stirred up, then the vital Spirit became to be sensible, which is in man a Sulphureous spirit, kindling the body by a heat, exsiccates the superfluity of the earth by the subtlety of its substance, and governs the body in a constant motion. For after the heat is gone, then coldness gets the dominion, the spirit of life being gone, no sensibleness felt in the pulse and arteries, and a dead body is found instead of life, at the departing of the warm spirit of Sulphur; rational men ought to take this mystery into consideration.

The two first elements, the matter and form, being apparent, and having gotten a nobility by the two last Elements by light, the *Microcosme* was not yet perfect, the Creator allotted further an increase to the seed of the earth, as well as he did to vegetables and animals. God allowed unto earth an imagination for all sorts of seeds, and to bring them forth after their several kinds. Then the earth was impregnated by imagination which God allotted, and the earth brought these seeds forth in man's presence, the heat digested them to maturity even till hitherto.

Matter and form of the *Microcosme* being extant, consisting of earth and water, then the Creator caused a life into them by an inbreathed warm air, heating the cold earthly substance, giving a heat unto life and motion, which was the Soul, which is the true Sulphur of Man, spiritual, incomprehensible: sensibly felt by its own operation. All this being finished, then God allowed an imagination unto good, in the perfect understanding of Man, that by his imagination he could judge of all the beasts, and impose on each a proper name, and by that imagination he learned to know his wife also, that she had flesh and bones of his body. Then man appeared perfect, and that matter was made into a shape, of a sensible body. This form being made alive by the Soul, had allowed further a subtle Spirit unto imagination and knowledge, which is an invisible, and incomprehensible form, like a work master, who frames all things in the mind, which has its habitation in the upper Region of the *Microcosme*, according to his volatility, and deserves the name of *Mercury*, or the invisible spirit of man's body. Form and matter is earthly, the life sticks in the motion, and the knowledge of every understanding unto good and bad stands in the sharp speculation of the *Microcosme*, the over plus found besides these three, nature casts off as *Cadave*, and is as a Monster, which by these three is found to be a separation, and a *Caput mort*.

If glorified *Elias* were present, and the *Astrals* could speak, and silent nature had a tongue to express hereof, then I needed not to bring in any further evidence to persuade the incredulous who considered not judiciously this my saying: for a man possessed by blindness cannot pass any judgement upon my writings: but understanding judge's impatience, and wisdom separates herself from folly by her own experience.

This vital spirit nourishes, feeds, and preserves himself by the Oleity of man's Sulphur, which is predominant in the blood and with, or through it does work in the whole body, that the substance may be perfect. This Vital spirit is Mercury, which is found in man, and is preserved by an Oleity of its likeness; besides these two Mercury and Sulphur, there is a third thing in man, namely Salt, which lies in the flesh, body, and bones.

The Salt ministers its noblest spirit for a nourishment unto the blood, which fatness is found therein by the taste, and disperses itself throughout the body, preserves man's body like a Balsome from putrefaction, and is as the hand and copulation, whereby Mercury, or the Vital Spirit continues longer with the Balsome in the flesh, and dwells together in one. For in the Salt there lies a spirit, which must protect all other Balsomes in their worth and dignity. The remainder found in the flesh, if these three be taken from it, is a dead thing, as I told you formerly, and is good for nothing, nor can it be used for anything.

As this Union, Dominion, and Government is in man, the like are in Metals, Minerals, and Vegetables, which make up their perfect body, do live, keep, and are preserved in the like manner, as man is. As the one follows upon the other in man, according to order in the like condition are other Animals after their kind and property. As a cow is an animal, her food, as grass, is Vegetable, this Vegetable by the heat of the cow's body is putrefied, in that putrefaction is made a separation, which is the key of all dissolutions and separations, separation being made, then goes the subtle spirit, the subtle Sulphur, and the subtle Salt of the Vegetable's substance of the grass into all Members of the whole body of the Cow, the spirit rules the beast, the sulphur nourishes it, and the salt preserves it.

This being done, then Nature distributes her gifts further, making a new separation; as of the superfluousness, which the cow does not assume by way of assimilation, and must part with it, and distributes the same further, and that is Milk, this Milk is an Animal substance, transmuted from the Vegetable. In this Milk is made another separation by fire, which must be kept gently. For the subtlest spirit of the Milk together with the Sulphur sublimes, is taken off, and turned into a coagulated fatness, which is Butter. The rest is separated by other means, and precipitated, and thereby is made another separation, this is a second coagulation, out of which men make their food: of the over plus, is made another separation by fire, not so fat as the former two: at last there remains an aquosity, and is of no great usefulness, because the spirit and its nutriment is taken from it by the several separations.

After this nature makes to a further putrefaction another and gross excretion of a Sulphureous and Salt substance, which generates afresh a living spirit, which is the excrement, this serves for the earth to be manured withal, making the earth fertile by its Sulphur and Salt, as being of a gross and fat substance, whereby new fruits are produced, here is another nutritive from an Animal into a Vegetable. This makes Wheat, and other Fruits, and Grains to grow, producing again a nutriment from the Vegetables unto Animals. Thus one nature does follow after the other, by vulgar people not so much comprehended, or searched into, not caring to learn Natures qualities so much, which makes such natural things seem to be incredible.

To return again to the structure of Man, the noblest spirit of life has its dominion and seat for the most part, and most strongly in the heart of man's body, as in the noblest part: and the Sulphur of man gives unto that spirit a nutriment, and spiritual access for its preservation by the air. For if air be taken away from man, then the spirit of life is choked up, departs invisibly, and death is at hand. The noble Salt spirit is a conserver of both, its noblest spirit penetrates throughout, the grossest matter of its Salt is cast into the bladder, and that has a spirit of a special operation. That which goes from the Salt through the bladder, is wrought upon by a heat, ministers a new access, or increase, so that this increase of Salt in man is inexhaustible, unless it dies quite, and the body be burned to ashes, and the remainder be extracted. As for example; Take the Salts from Minerals, let these grow again, coagulate, and extract the Salt again by water, the like is seen in nitrous earths also, and other common Salts, and there needs not to quote any further examples.

The spirit of life has its process into other Members from the heart, into bones, arms, and the rest of the body that are stirring; In diseases and symptoms he is weak, and man by reason of such symptoms, cannot perform his business in that full strength as at other times, when he is in health, feeds and cherishes his body with vegetable spirits, which come from feeding bread, meats, and drinking of wines, then his body grows stronger, and his Vital spirit grows potent by such nutriments, in his superfluity disperses himself into all Members, and shows his operation. If the heart grows faint, then is it a sign that the Vital spirit is not nourished, upon which there ensue speedily deadly diseases; because that fire is not at liberty, and falls into an extinction, or choking.

The fire in the heart, and the natural heat is preserved, and supported by the air, of that air the Lungs stand most in need of: The liver also must have air, else it cannot laugh: The Spleen must have air, else it will be oppressed, with stitching's and great pains; The true seat for the most part of the air is necessary for the Lungs, if these fall into any weakness, the cause thereof is, because the Salt does not show its true, and meet help, and must go into rottenness, casting up blood and matter: Then there is at hand a corruption of the air, from which the Vital spirit cannot find any true nourishment, but must be starved, because the Salt does not affect its conserving quality, the Sulphur, and the increase of the nourishment is obstructed, and is not perfect, whereby are caused Consumptions, withering of the body, consuming of the flesh, and exsiccation of the blood, and of the marrow. The substance of Salt, or the Salt spirit, which preserves the body, has its seat for the most part in the bladder, where all humidities have their issue, the remaining gross Salt is separated, and excerned by Urine, as you heard already. I repeat it here again for that end, because the most noble spirit, which does preserve man, does copulate, and make friendship with the Vital spirit and its nutriment, which is Sulphur, and so they make the body perfect, and if any infirmity be incident, either from the operation, or defective quality of the Stars, or from a disorderly life in eating and drinking, and many other inconveniences, and any corruption be present, then nature is not in her perfect condition. Here the knowing Physician must enquire into the cause, from which of these three the Symptom does arise, and cure the same with convenient remedies, and not with any contrary Medicines: as heat must be cured with heat, cold with cold, prickling with prickling: for one heat draws the other, one cold draws another, even as Iron is drawn by a Magnet; and so prickling simples may cure stitching diseases, and poisonous Minerals can heal, and bring to right poisonous Symptoms, if they be duly and well prepared. And although sometimes externally a cooler be supplied, however I speak as a Philosopher, and one that is experienced in nature, that like must be cured and expelled fundamentally with the like, otherwise true Medicaments are not applied, and the Physicians deal not really in their profession. He that is not fundamentally learned herein, or does not observe these things, he is not a true Physician, neither can he really say that he has learned any truth in Physick, because he is not able to discern cold and heat, dry and wet, for knowledge and experience, and a fundamental inquiry into natures mysteries make a good Physician next the knowledge of the Creator, from whom all, and every wisdom does descend, and is the Author of the beginning, middle, and end thereof.

Next the invocation of the Creator, there follow natural means and Medicaments, as they are found in themselves in their highest degree; I make no further mention here of other Animals: Metals and Minerals follow next, for in Gold, Silver, and other Metals, even to the seventh and last Planet, are hid excellent things,

Mercury being predominant in all, in some more than in others, and Minerals also are not without their virtuous Medicines, and the former seven Planets were in their beginning only Minerals.

The tincture of *Sol* together with the potable Gold and Silver are of great efficacy. Mercury rules Microcosme: that, which is found in the best Metals, and most precious stones may be drawn also, if need be from Minerals. For perfect Metals are grown, or have their descent from Minerals, as from Vitriol, Antimony, and the like. Vitriol is Sulphur, Antimony is Mercury, the Salt which is the copula, or binder, is found in both, if these are made fix, are like unto the best Metals, for they are generated by them: Minerals come from the three principles, as well as Metals: The three principles come from their *prima materia*, called *primum Ens*, which is nothing else but a watery substance found dry, is not likened to any matter which is grown, and is preserved by the four Elements, and these are cherished, or nourished by Astrals. The Creator has ordained all these out of a naught, because man should not gaze only upon earthly matters, but consider heavenly ones also and ought to know things supernatural, that faith may over-top the rest, and have the prerogative in things seen and felt, and be preserved therein.

If Physicians do not understand these things, they ought not to be held for Physicians, for the knowledge of God and Nature make a Physician, as I told of it formerly, and not great prating without true knowledge; Good writings of expert men may conduce somewhat hereunto.

In brief, humane reason in Physicians is not able to comprehend sufficiently, much less are they able to decide, fathom, and fully learn, what manner of Medicaments there may be made of the *Microcosme*, for he contains a perfect Medicine for all diseases, like with like must be expelled and cured. *Mercury* of the *Microcosme* is a living, incomprehensible, and volatile spirit, as I have told.

Man's Balsom dries up a Dropsie, and the clarified Salt of it cures Consumptions; in Epileptic fits it does excellent well, and being prepared into a fragrant spirit, all corrosiveness being taken from it, is nothing inferior unto *aurum potable*, to preserve man's health; it is very excellent for curing Leprosie: Passing by such diseases, which are of a lower nature and degree, it breaks the stone in the bladder, and cures all Salt Rheums, if the artist prepares it well, and knows how to make use of it afterward.

Thus I close with the *Microcosme*, contained in few lines, much more could be spoken of this matter, or form, mobility and imagination, how they were brought unto perfection. For if these stand together in a true middle, will make up a sweet Harmony, for without the matter, or form of the body, without the moving of the powers, and defect of perfect thoughts *Orpheus* will not please the *Dolphin* with any harmonious melody: as it is with man, so it is with Metals: Mercury is the *mobile* in

Gold, if the body be anatomized: Sulphur is hot, being driven from a Mineral and fixed, dries the phlegmatic *Lune*, warms her, makes her Soul equal unto himself. In the matter and form there lies a Salt, which affords the coagulation of the body: the remainder in the Gold put away, for separation will afford a further revelation.

Vegetables also show the form of their three principles, the visible matter contains the Vegetable Salt, which is its conservative, the fragrancy of the vegetable is the Balsam, which ministers a nourishment unto its perfect growth, the odor, or smell of any Herb is of a volatile quality, and spiritual, and the spirit for the most part shows itself in the fragrancy, and penetrates the Balsam, and its odor, be it pleasant, or not, is the essence, whereby men in their senses learn the condition and properties of Vegetables. For other things I have written, I praise the Lord, which dwells on high. Thus closing I wish to everyone the grace and blessing of God the Creator of all Creatures, that they may be blessed, wise, and rich, both in this temporal and corruptible World, and in the other World attain to an eternal bliss. AMEN.

Of the Mystery of the Microcosme, its Medicinal Parts
belonging unto Man,
written by
Basilius Valentinus.

To make use of the heavenly Revelation, about two Luminaries, and of the mysteries of the whole Medicine, which lies in that marvelous instrument of the *Microcosme*, within and without, that is, in the body, and without, as ordinary Wounds, Sores, Ulcers, that have their cause from within, have their descent from one root, however, must be severally prepared and dressed. For that within is not like to that which is without, in respect of their operation; but in respect of their form and matter they are under one judgement. And that I may rightly inform my fellow Christian, I must needs acknowledge and confess, that there are two medicines, which heal all diseases and symptoms, be they whatsoever, and are made of one, the one is called PHALAIA, and is for inward use, the other is called ASA, is for external cures; both may be called to be only one, they differ only in their preparation; how both must be brought to their operative quality, the way unto it is showed in my *Manuals*. For they must first be rightly known, and their nature must be searched into. Their matter is One, which by that expression I purposely hold it forth, lest it should be made too common. I after the manner of Ancient Philosophers before me, hiding secrets under dark sentences, hoping by the prayers of others to have their Souls saved, and received into that Garden, in which our first Parents were Created.

Note, both Medicines are made of one matter, as I have already informed you. If used inwardly, it takes away all manner of infirmities: the matter is putrefied, separated, and in a Spagyrick way purged in the best manner, and brought to a Medicine of the highest degree, by fixing its own nature, which must be brought to pass in the fire. Its former poisonous volatile quality must be rectified, by being prepared to an everlasting fixedness, which expels, purges, and rectifies all malignant spirits, that a good nature may live quietly in a pure habitation. For this prepared Medicine, keeps that course, wherever it needs with any malignity, it will be revenged on it, and strives to expel it, and will solely keep possession there: For she cannot endure any contrary things about her, which are defiled with the least impurity.

PHALAIA is the Universal Medicine to be used inwardly, and ASA is the Universal remedy for outward uses: it purifies man's blood, takes away all impurity, strengthens the brain, heart, stomach, and all other parts, causes good blood, strengthens the memory, repairs the defects, which are befallen the three principles, restores all lost things: It is the very Key, whereby the body is opened: For it chases away Leprosie, Consumptions, Gout, Dropsy and all other diseases generally: For no sinful creature is fallen so totally, but she may have a comfort unto salvation in a spiritual way, and a Medicine unto health, appointed thereunto by the Creator, which is had if Nature be anatomized by an expert Artist; to be prepared for that use.

Here I speak of such diseases, which by some are called incurable: for ordinary diseases, there are ordinary means which here are not mentioned, the uses of them are mentioned. in a special Treatise.

But of my PHALAIA I say this much, according to my long experience, that nothing can conceal or hide itself from her, being a penetrating searcher into all infirmities, she penetrates the body spiritually like a fume, penetrates the Arteries, Muscles, and all the parts of the body like a Balsam, restores strength which was left by her Salt Spirit. Further, I cannot speak in the praising of my PHALAIA, she being a praise to such that make use of her. He that gets this PHALAIA rightly, to him is she sufficient for to cure all diseases. No tongue can express and set forth fully her virtues.

As diseases do differ, which are incident unto the body, so there are means for their cure: But this Medicine cures all diseases in general, being of a heavenly sidereal quality, descended from the Elements, and generated by the three principles, coming from the very heart of its Center of the whole circumference of the Globe performs all, affording to the *Microcosme* a perfect medicine found so really, according as the name imports her virtues, but if rightly made and prepared, the use of it will prove it sufficiently.

ASA is found in the operation, for external Symptoms, as old Ulcers, Fistulas, Cancers, which made many Chirurgions doubt whether ever they may be cured, but this ASA has made the cure: it consumes all bad blood, which was fallen into corruption, and may be inwardly used, because it will then exsiccate, and dry up the fountains, from whence springs all manner of Sores, Fistulaes, Cancers, Wolves, *noli me tangere*, running Legs, Worms, and the like, be it on what part of the body, where Plasters, Poultice's, and the like cannot help, and are not strong enough, this alone will do it. For fresh wounds, be they made by stabbing, cutting, slashing, it needs not to be administered, being too strong for such wounds, gentler means are fitter for them, Balsams, Oils, Plasters, may heal them, either outwardly, or inwardly; Powders and Potions may be prepared. Symptoms in wounds, having their causes from within, must be cured by searching into them, and the means for their cure must be prepared of that strength, that they may reach home. As in this matter, things must be united, and be taken from the generation of ABIHAIL, being joined in their principles of the first essence, by natures means, is brought to the highest perfection, whereby such Sores, Ulcers, &c. are fundamentally cured. For ordinary wounds, there is no need of it, if no Symptoms are at hand, and the party endangered, a Balsam only will perform the deed, mollifying the flesh, and Nature will further, and promote the cure.

Be thankful next God to me, that has taught you inward and outward Medicines, and are such, which others before me have concealed, they can cure fundamentally any Symptoms, be they within, or without on the body, such virtues are not found in outlandish woods, drugs, or herbs: Foreigners have their proper climates, under which they have convenient Physical Vegetables: Our climate affords unto us proper Medicinal Vegetables, Animals, a:rd Minerals for our constitution; only Doctors are not expert to make their Medicinal preparation out of them. I hold with my Physick PHALAIA and ASA of one name in their descent, whereby Nature has made me to be a Physician; it keeps good to the last, preserves man in health and strength all the time the Creator has appointed for him: virtue it has showed triumphantly in many parties, obtaining victory against all its enemies, and it was apparent to the world, that these two Medicines PHALAIA and ASA of one kind, and of one matter made and prepared: and is found daily, that in the generality they can set all into a perfect degree, as being descended from the Center, can preserve the Center as the Root, and can bring things to right within and without, tending to that end, for which it is prepared.

Thus, I wish the reader, to whom I faithfully intimated the Manuals of it, prosperity and success in the preparation of it, that it may be unto his health: The work will praise the Master, upon my Oath I further inform you this, that four things are required to make a perfect Philosopher, and true Physician.

First, he must be importunate and fervent in his devotion to God, as the highest heavenly Physician, to ask of him grace, wisdom, understanding, and his blessing upon his undertakings, that it may appear unto the world, that God grants things for the good of men, that he may be praised and magnified for such benefits: and is to show himself in his life and conversation godly and honest.

Secondly, a Physician ought to know the diseases, and to distinguish the one from the other, and what proper remedies he is to use against these diseases: For without the knowledge of diseases a physician is not perfect: man's complexion must be discerned, the cause of the disease searched into, and the means well considered, that no contraries be applied, whereby further troubles are caused: proper remedies fitting the diseases must be applied, that restitution be made unto former health.

Thirdly, it is requisite, that he read frequently the writings of Ancient Philosophers, and read them over and over, and take notice in what they do concur and agree; and where they aim all at one mark, then he that has understanding will discern the good from the bad, sophistry from truth: The ancients knew many good things, for mine own part I must confess, that I borrowed the foundation of my knowled.ge from them, which made me to lay it to heart, and am thereby moved to leave for others also a Cornerstone, that truth may further be confirmed, and grounds of it made easier, clearer, plainer, and more manifest by a further knowledge of my writing.

Fourthly, a Philosopher must learn to Anatomize things in Nature, to know what they contain within and without, to separate the poison from the Medicinal quality. Hereunto belong several Manuals, how to dissolve, separate, exalt; and prepare fully Metals, Minerals, Vegetables, and Animals. He that has learned all these, he may discourse wisely of things, confirm their grounds in truth; but others, which are ashamed to work herein, cannot glory in any truth: because by the receipts found in other men's writings, not endeavoring to learn more in their own experience; I am not ashamed to learn daily, because Nature is round and endless, cannot be comprehended fully, by reason of the shortness of man's life, and none can say that there is nothing left more for him to learn. No such matter. Thus you see, that God's blessing must be obtained by fervent and frequent praying unto God, the causes of diseases must be known, their cure must be ordered according to the direction of Philosophic writings, adjoining an experimental knowledge thereunto. He that does, and knows these four things may glory in his ways, confirming things in deeds, and not to exercise a trade upon other men's receipts.

My Medicaments, if well prepared and duly used, will by Gods help make known, that they received their strength from God, the marvelous Creator to perform these things, which ignorant, and men of little faith cannot comprehend: by daily experience faith gets strength, that man may praise the highest, who has put such virtues into natural things, for the which mortals are not able to return sufficient thanks. As much as lies in my power I will praise the Lord day and night, and is not possible to require him in any other way. At the closing observe thus much: in school long discourses are made of the three principles of all things, of the matter of heaven, what it is made of, and on what the earth does rest, how the Elements were made, and of the beginning of the firmament, and of several opinions they are about the original causes of Metals, Minerals, Vegetables, of their qualities and properties, of the original of man, and of other Animals, searching in their conceits into their lives, virtues, *&c.* But my Son hearken unto me, and take notice of what I say; all their pretended sayings are a mere nothing, they speak ignorantly without any certainty; because they have no experimental knowledge, having laid no foundation, nor have they learned any true decision in their demonstrations: Thoughts pay no Custom, or Toll, they fly into heaven, descend to the nethermost parts of the earth, if experience and their thought do not concur, then their thoughts are found a mere opinion, then they must confess, I did not think it could be so! Man's thoughts are fitly compared with a dream, because nothing follows upon an imagination: Natures secrets must be studied experimentally. If Artists or Mechanicks would imagine to work such, or such things, be it Watches, or other curious Metalline works, but does not invent fit instruments, whereby to make that work they have in their fancy, what can they produce by that imagination? An empty opinion, and no Art. So, in the knowledge of Natural things, their Secrets require a greater exactness to be searched into, which to lazy inexpert men seem strange and impossible. I tell you there is required an exact diligence to find that, which lies hid in them, it must be done by separation. Nature must be anatomized, good and bad in it must be discerned, what is contained in each in its Center, for the general, and what comes from it in particular.

Therefore, the *Macrocosme* and *Microcosme*, yea, the things which grow and are found therein, are compared to a round Circle, in whose middle there is a Center, let the Circle be turned which way it will, it keeps round every way, and its Center stays unremoved. A Philosopher must also know rightly the Center of each matter, which must stand unremoved in every substance, but the substance may be turned any way he pleases, and make of it several forms, according as it received its power from above. I speak now to be taken notice of thus: I take in hand any natural thing, dissolve, or open it by a Key, which is the means of the unfolding, and search therein by a fire's proof, which is the master of all proofs, what may be made of it: Here I find as many wonders and qualities, which I never thought of, much less had I experience of.

Of natural things are made Powders, Oils, Water, Salt, Volatile Spirits, and Fumes: In these preparations are beheld wonders upon wonders, witness the distillations, digestions, and putrefactions. There are found and seen many spiritual and corporeal colors, which appear black, grey, white, blue, green, yellow, red, azure color, with a reflexion of all manner of in sprinkled colors, which cannot well be described, and inexpert men hardly believe it. From these preparations are several qualities felt, the one is corrosive and sharp, the other is pleasant and mild, the one is sour, the other is sweet, according as they are prepared, so they yield good or bad, poison or physic: for a good thing can be made the worst poison, and the worst poison can be turned into the best Medicine: which is not so great a marvel, because all lies in the preparation of things: though every one cannot conceive of it, yet it is so, and will be a truth forever, because nature has manifested herself thus by experience.

A blind man cannot tell how the inward parts of man's body are conditioned, but the seeing Physician, who anatomizes the body, he can judge of the situation of the Heart, Brain, Liver, Lungs, Reins, Bladder, of the Entrails, and of all the Veins, and knows in what form and condition they are. But before he has made this anatomy, all these were hidden from him, a Miner who seeks for Ores, he does not know what riches he may expect from Metals, unless he opens the Ore, and so fine it: what he finds in it by fire, then he may know really in his calculation, what riches he may expect from it. So, other things must be proceeded in, which true Naturalists will endeavor to do, and not prate of things only without experimental knowledge, disputing of colors with the blind man, learn to know the ground with your own eyes and hands, which Nature hides within her, then you may speak wisely of them with good reason, and you may build upon an invincible Rock. If you do not so, then you are but a fantastic prater, whose discourse is grounded on sand without experience, and is soon shaken by every wind, and ruined in the end. The ground of this knowledge must be learned as you heard, by anatomizing and separating of things, which by distillation is made known: where every Element is separated apart, there it will be made known what is cold, or moist, warm, or day. There you learn to know the three principles, how the spirit is separated from the body, and how the Oil is separated from the water, and how the Salt is drawn from the *Caput mort* of each matter, and is reduced again into a spirit, and how these three are afterward joined again, and by fire are brought into one body. Further is here learned, how each after its separation, and afterward in a conjunction may profitably and safely be used for their several uses they are prepared for: all which must be done by a *medium*. At the first Creation man is earthly and gross, but his Soul, Spirit, and Body, being separated by death, putrefies underground, and when the Highest comes to judgement, he is raised again, his Body, Soul, and Spirit come together, according to Faith and Scripture; that body is no more earthly, as it was formerly, but is found heavenly and clarified, glittering as the Stars in the East, and like the Sun is seen, when all the clouds are past. So, it is here, when earthiness is broken, divided, and separated, then the three principles of the dead substance are made apparent, the dead one is forsaken, the living power comes to her perfection, because her obstruction is laid aside, that the virtue in the operation maybe manifested; In this separation and manifestation is then known what these three principles are, which are so much discoursed of, namely Mercury, Sulphur, and Salt, according to the condition of the Subject. He that does not think it to be true, let him go to the end of the World, **where** he shall feel all, what in his dumb capacity he could not comprehend: If anyone should intend to teach me any other with a prolixity of words, he may fill me with words, but he must prove it really also, for without that I am not bound to believe his words, but desire some sign, as *Thomas* one of the Twelve, who looked for an Ocular demonstration: I might have left out *Thomas*, but being there is a *Gulf* between a spiritual and worldly unbeliever, I gave liberty to my mind to speak it, for there is a great difference in heavenly and worldly matters, touching faith and things comprehensible, and there is that difference found also in sidereal and earthly things: For sidereal things are comprehended by sharp imagination, and Arithmetic rules, but to the finding out of earthly things there belongs speculation and separation: With speculation must be joined an intention, and an apprehension is annexed to speculation, the former is done spiritually, because the spirit of man does not rest, desires to apprehend more qualities of the spirit in things natural: every spirit still draws its like: the rest is earthy: for an earthy body separates by manuals the earthly body from the spiritual part, and so the one may be discerned before and from the other. Whereas the Soul in both shows herself really, therefore is she in all really, for she ties the heavenly and earthy together like a bond, but when the heavenly is separated from the earthy, that the soul also must forsake her body, then you have separated and received the three as apart, which after a true knowledge and conjunction can afford such a triumphing and clarified body which is found in a better degree of many thousand time because the grossest is laid aside from the earthy.

For when heaven and earth come to be refined by the great Creator, then the greatest part will be consumed by fire, and by that purging it will be exalted to the same degree with the heavenly, and set into the same line, for each all is created by one, each all is ordained by one, and though through sin by one man all was corrupted unto death, yet all is by one brought to a better State of life: and the only Creator intends to judge all by fire, and all must again become one, which will be that heavenly essence, to which the earthly gave way by means of the fire: the eternal glory leaving a room for devil and death, from whence they shall look on the elect, admiring the great Majesty and glory of God, which in a divine essence of three distinct persons is all in all: and has created all.

Thus the three persons in the deity have held forth in us three invisible essences, giving thereby to understand by an unsearchable wisdom, what their creature and order is: we men are too weak to come higher; God is and will be God, and we men must be content with such gifts afforded unto us: hereafter shall be accomplished that which is prophesied of by the Prophets and Apostles, and now are conceived of only by way of faith, therefore we ought now to be converted, what by Nature is intimated in a visible way: other things incomprehensible unto us, and matters of faith, will appear better to be understood at the end of the world, God grant unto us all a true knowledge of temporal goods and of the eternal.

At the closing of this I say, that this is the whole Art and whole foundation of all the Philosophic speech in which is that sought, which many desire, taking great pains, and making great expenses, namely to get wisdom and judgement, a long life, health and riches of this world comprehended in a few words; as for example. First you must know, that I will show unto you such an example of the Animals, which in the appearance is a mean and poor one, but of a mighty consequence, if rightly considered. The Hen lays an egg, the same egg is brought by heat to a hardness or coagulation; by a further heat it is brought to a putrefaction, where it is corrupted: in this putrefaction, the egg receives a new *Genus*, wherein is raised a new life, and a Chicken is hatched. This Chicken being perfect, the shell opens making way for the Chicken to creep forth, this Chicken coming to a further ripeness and age, increases further in her kind. Thus, Nature furthers her own kind, and augments *Usque ad infinitum*. True, the egg is not *prima materia* of the Cock or Hen; but the *prima materia* of their flesh is the first seed, out of which the egg is gone into a form, which by the equal nature of the motion of both is driven together and united, from thence by a further heat it went to a putrefaction, from thence into a new birth, which new birth still propagates and increases.

So, it is with man, for one man alone cannot produce a new birth, unless both seeds of male and female be united, for after this conjunction through Nutriment of the body, and continued natural heat of these two seeds, which in the Center are known for one Nature, get a new life, and more men are begotten, which propagate further by their Seed, by this means the whole world is filled with men. This seed of man is the noblest subtlest blood of a white quality, in which dwells the vital Spirit, which is driven together by motion. If these seeds of both kinds by their desire of lust are together united, and their Natures be not corrupted, or else are contrary one to another, then there is preserved a life by a heat, and brought to perfection in the mother's womb, and another man is brought forth. Thus, much be spoken of the Seed of Animals.

The vegetable seed is made palpable and visible which from each kind of herb is separated and propagated in the earth for an increase, which seed must first putrefy in the earth, and then must be nourished by a temperate moisture, at last this seed by a convenient warm air is brought to a perfection, thus Vegetables are increased, and in their kind preserved: but the first beginning of a vegetable seed is a spiritual essence or astral influence, whereby in the earth was gotten an imagination, and became impregnated with a matter, out of which by the help of the Elements it came to be something: what form of seed the earth was desirous of after the heavenly impression, that form it received first, and brought it to a kind, which brings a further increase by its palpable seed in the generation, hereby man may try his further skill: but he is not able to create a new seed, as Nature does by an influence from above only he is able to increase a formed seed.

Of Metals and Minerals, I inform you this, that there is one only Almighty Being, which is from eternity, and abides unto eternity, which is the Creator of heaven and earth, namely the eternal Deity in three distinct persons, which three in the Deity are a perfect divine Being: and though I confess and acknowledge these three persons, yet I confess only one God in one Being. This I do now speak as a Type of the first seed of the three principles; that the first beginning, to beget Metalline seeds is wrought in the earth by a sidereal impression, which quality presses from above into the nether as in the belly of the earth, and works continually a heat therein, with the help of the Elements; for both must be together: the earthly affords an imagination, that the earth is fitted for conception and is impregnated, the Elements nourish and feed this fruit, bring it on by a continued hot quality into perfection, the earthy substance affords a form thereunto; thus at the beginning the Metalline and Mineral seed is effected namely by an astral imagination, Elemental operation, and terrestrial form: The astral is heavenly, the Elementary is spiritual, and the earthy is corporeal, these three make of their first Center the first essence of the Metalline seed, which Philosophers have further searched into, that out of this essence there is become a form of a Metalline matter, palpably joined together of three, of a Metalline Sulphur heavenly, a Metalline Mercury spiritual, and a Metalline Salt bodily, which three are found at the opening of Metals: For Metals and Minerals must be broken and opened: Minerals are of the same sanguinity, of the sane quality and nature, as Metals are, only they are not sufficiently ripened unto coagulation, and may be acknowledged for unripe Metals, for the spirit in them is found as mightily Metalline as it is in the most perfect Metals. For Metals may be destroyed and easily reduced unto Minerals, and of Minerals are prepared Medicaments, which ripen and transmute Metals, which must be noted: and it is done, when Spirit, Soul, and Body are separated and purely re-united. The remaining terrestrity being put off, then follows a perfect birth, and the perfect ripening by heat performs her office, that Spirit, Soul, and Body at the beginning in their first seed have been a heavenly water, which begot these three, out of which three is become a Metalline Sulphur, a

Metalline Mercury, and a Metalline Salt, these in their conjunction made a fix, visible, palpable body; first began a Mineral one, then a Metalline by an astral imagination, digested and ripened by the Elements, and an earthly substance are made formal and material.

Now when these bodies of Minerals and Metals are reduced to their first beginning, then the heavenly seed does appear and is spiritual, which spiritual must become an earthy one by the copulation of the Soul, which is the *medium* and middle bond of their Union to make a Medicine out of it, whereby is obtained health, long life, wisdom, riches in this mortal life: this is the true sperm of Philosophers, long sought after, but not known: whose light was desired of many to be seen, and is even the first matter, which lies open before the eyes of all the world, few men know, it is found visibly in all places, namely Mercury, Sulphur, and Salt, and a Mineral water or Metalline liquor, as the Center, separated from its form, and made by these three principles.

The Heavenly Physician, the eternal Creator and inexhaustible fountain of Grace, and the Father of all wisdom, Father, Son and Holy Ghost in one Deity, teach us to know really in a due gratefulness his wondrous works, and make us co-heirs of his everlasting goods, that we after a temporal revelation may in a true light seek the heavenly treasures, and may possess them eternally with all the elects, where there is unspeakable glory without end, which is attained unto by faith in our Savior by bringing forth good fruits, by loving of our neighbors, and helping the needy, which must be made evident with an unblameable life, and due obedience to God. Amen.

FINIS.

TWO TREATISES

OF THE MOST EMINENT AND INCOMPARABLE PHILOSOPHER

BASIL VALENTINE

FRIAR OF THE ORDER OF THE BENEDICTS.

THE FIRST

Whereof declared his Manual Operations, how he has made and prepared his Secret Medicines; the Stone Ignis out of Antimony, and last of all the Philosophers Stone.

THE SECOND

Discovering things Natural and Supernatural, as also the First Tincture, Root, and Spirit of Metals and Minerals, how they are conceived, ripened, brought forth, and augmented.

Printed heretofore in the German language, and now for the good and benefit of the English Nation, translated into English.

The Epistle to the Reader.

Courteous Reader,

You have here two excellent treatises of that incomparably experienced Philosopher Basil Valentine: The first whereof, *viz.* his Manual Operations, is one of the most perspicuous and clearest of all his Books, which ever he left behind, or were published: If you do rightly prepare those Medicines, and administer them to the Patient, you will find that I have communicated to you a precious Jewel. And moreover, that you might not want those Noble Medicines, in case you should want either skill, or time, and leisure to prepare them. I am resolved with the assistance of Almighty God, to prepare always some of the Magistery of Antimony of our Philosopher, which he teaches you to make out of Mercury and Antimony, and has been found by me and many others a most excellent Medicine in many desperate distempers, and do intend to leave it with the Stationer, Master Edward Brewster, at the Crane in Paul's Church-yard, that has been of the charges of the printing of this Book; so that you may have it there, whensoever you have occasion for it, and at so reasonable a price, as you yourself will hardly be able to prepare it at a cheaper rate. But you are to know, that I have exalted this Medicine, and prepared it much better than our Philosopher, because I have fermented the same with a Volatile Essence of Gold, and then fixed them together, so that I do account this *Magisterium Antimonio-Solare* may deservedly be esteemed a *Panacæa*. But because the Stone does require a Medicine of another nature, I have added our Philosophers Medicine against the Stone, which you may have likewise at the aforesaid Stationer-shop. Nevertheless, I have added some things which in my practice I have found to be extraordinary good against the Stone, to the things which our Philosopher makes use of. And having thus exalted this Medicine, I do not doubt but you will confess, after you have used it, that you never have found a more powerful Medicine against the Stone. And because you want many a time a very good Purge, which yet our Philosopher has not in this Book of his, I have likewise provided for you a gentle, yet excellent Purge, made out of the above said *Magisterium Antimonio-Solare*, of which I am sure, and you will find it by experience, that it purges the body very gently of all noxious humors, of what quality soever they may be, so that it may rightly be called a *Purgans Universale*. And thus I hope these three Medicines will serve you, if not absolutely for all distempers, yet for the greatest part of them, as well in *Chyrurgia* as *Medicina*; and you will not be necessitated to follow and to make use of a great many uncertain Remedies and Recipes.

THE MANUAL OPERATIONS
Of
Basil Valentine,

Whereby he has prepared his Medicines.

In the name of the Eternal Trinity, God the Father, God the Son, and God the Holy Ghost, I *Basil Valentine* do here set down those Manual Operations, whereby I have prepared my following Medicines, which by Gods assistance have made me a successful and scarce ever failing Physician.

But before I set forth those Medicines, I must here remember, as many authors have done before me, which I well approve of: That the Ancient Searchers of Nature, who have lived long ago much before me, have written of a Bird, named by them *Phœnix*, and is still at this very time called so. Not that such a Bird is existent, or to be found in the whole world, that flies from one place to another, looking after her meat and breeding her young, for indeed there is no such thing. But the *Phœnix* is a fictitious Bird, which is never consumed in the fire, but renews her age therein, and her kind is raised by the fire, so that she lasts to the end of the world. Thus, likewise it is to be understood of Medicines, which must cure, and by rooting out consume fixed diseases, that they must be prepared so as to be fixed, before they can dispel fixed things. For nothing that is slight, or feeble, and weak, can resist that which is strong, but the strong must be cast out by a stronger. Therefore, the Ancients have invented this Bird, and compared her with our true Stone, being the Universal Medicine of the World. Besides this Universal Medicine, there are prepared many other Medicines, which indeed do not consume diseases universally, as our Stone does, but do work particularly every one curing certain diseases, whereto they are ordained by the Most High from the beginning at the Creation for the good of Mankind, which are to be further prepared and perfected by the Physician. For hot distempers require their own Physick; cold distempers, having their original from cold, do likewise require a proper remedy. The like does mixed distempers, which are of a middle nature. All this must the skillful physician know and understand, if so be he intends to gain credit by his Art: On the contrary, without such knowledge he will not in all his lifetime gain any esteem, but loose his credit and reputation. For every Physician, must consider, that there is a great difference betwixt those diseases, which have fully possessed, and clearly overspread the whole body, as the Leprosy, and such like; and those distempers which have but taken up their lodging in a man's body, as a traveler does his at an Inn; such are the several sorts of fevers, and other like maladies. Therefore, every single medicine must be directed and applied to the disease, to which it is proper and deputed. In like manner, External Distempers have their peculiar Natures, and a certain difference must be made betwixt them. For in old, lasting, and spreading Ulcers and Sores, which arise from within, another way of cure must be used, then in healing of simple, common ,green wounds, outwardly made upon the flesh, which may be well cured on-y by outward application of certain Ointments, Plasters, Salves, Herbs, Balsams, and Oils: And (except some singular accidents by the influence of the Stars of Heaven, should require the preparation of some healing Drinks to be inwardly taken) those aforesaid Medicines may be sufficient alone to cure any green wounds, without further addition of any other inward means. But this cannot be in old Ulcers, which have their original from within. For their original being internal, there must be internal Physick likewise administered, whereby those humors, which keep them open, may be dried up, and their issuing forth stopped. But maybe some Physicians say; how can we remember all this? This would cost too much labor, and much time will be spent in finding it out, and our life is too short, death will prevent us herein. A Physician ought indeed to know it, if so be he will be perfect, and discharge his office and calling before God and the World conscientiously, that the account of his stewardship may not send him to Hell. For it is not enough, that a searcher of the Secrets of Nature says: The earth is adorned with many sweet and delicate Flowers of all manner of colors, and that the Birds of the Air are beautified with several colors and pleasant feathers. This is not enough to make one a Philosopher, or searcher of Natures *Arcana*; because every Clown may behold the variety of colors in Flowers, blue and all sorts of mixtures. But when the ignorant Fellow by further enquiry is required to give an account of the original of all such colors, how those colors arise, and how they are driven out by Nature; he is then as learned a Master as Doctor *Coxcombe*, who was to taste some broth, whether it was salted sufficiently. Therefore, something more is required to learn to know everything, and to search into hidden secrets. For a searcher of Nature must know more than a silly Country-man, who only beholds the colors, which everyone may do, since they are exposed to everyone's view, but he must look back, and by serious speculations search and enquire, how those many colors visibly shining in Animals and Vegetables, are likewise set forth and do appear in Metals. If he finds and discerns this, he is then a true searcher of Nature indeed: but without it, he is no more such a one, than other un-experienced Country-men. I will speak no more of this, but hereafter I must tell you, that all Natural diseases, External and Internal are caused by two things, to wit, either by an earthly and grosser medium, as by inordinate or superfluous eating and drinking, or too much care, fear, watching, taking colds, and the like; or else by more spiritual and heavenly influences, as when Elements become infected, polluted and poisoned, whereby they produce many and manifold distempers in the lesser world. The first is more corporeal, the other spiritual, for it is produced after a kind of spiritual manner. The bodily distempers have their seat in the blood and stomach, from whence they work into the other parts, and cause pain: Such distempers may be cured well enough by more earthly and gross ways, as by purging and letting of blood. But as concerning spiritual diseases, wrought by malignant influences of the Stars, they are not removed by corporeal and bodily remedies, for they are much too weak: But it is to be observed, that if such a spiritual distemper have taken deep root, spiritual Medicines are to be applied, which are of that nature, that though they do look like corporeal ones, yet are they so prepared, that like a Volatile Spirit they penetrate the whole body, to sweep away all morbific matter, which no Medicine is able to perform, which not being separated from, lies as yet hidden in its gross body. What other distempers soever there be, that do not derive their original from these two natural causes above said, they are not Accidents of Nature, neither can they be accounted natural, but must be judged to have been wrought by Witchcraft, which cannot be cured as other distempers, and with such remedies: but if any desires to be freed of them, a Magical cure must be used, nevertheless such Magick must be followed, as is not against Nature, but rather does agree with it, and which does not dishonor our Savior, nor endanger our Souls, but work in a way suiting to Nature. There might be much said, and much written of these things, but I hold it needless for some pregnant reason, which I do keep to myself. For herein neither the wit of *Athens* can assist me, nor the power of the *Romans*; neither can the riches of *Crœsus* or *Abasuerus* pay me for what I keep back in those points of Magick, which I hope is prudently done of me. But if *Uranius* the Father of *Saturn* were yet alive, he would perhaps keep his residence in *Iron mines*, that so, besides the *Haruspex*, like a Cunning man he might give an account, whether *Haliætus* the *Sea-eagle* is gone, and where *Alcarmespan* the little *Saffran-worm* makes his Crimson.

This may suffice for a Preamble to my Manual Operations, wherein I have given you this account, which will be rightly understood, when those Manual Operations of mine are industriously and vigorously practiced, and his desire accomplished, and then the Eyes will be opened to behold, what now the Ears let in, *viz.* where virtue lies buried, and Truth overcomes falsehood.

The Creator of Heaven and Earth, the Son of God our Redeemer, the Holy Ghost our Comforter, who has sanctified us, be pleased to assist me, that I may successfully finish and put a period to this design of mine, to the glory of God, to the comfort and profit of my Neighbor, and to the promoting of the Salvation of my own Soul.

I will now first and in the beginning, write of some preparations of Minerals, and discover, how I have prepared them in the fear and by the assistance of God, and found them helpful in the greatest distempers. The universal shall follow hereafter.

Of Vitriol and its Preparation
as also of its Power and Virtue.

Take good *Hungarian* Vitriol, calcine it, till it be of a yellowish color, and no higher. Grind this calcined Vitriol small, put it into a distilling Vessel of Glass, with a long neck, well luted, *luto sapientiæ*. Put thereto a large receiver, and begin to distill day and night with a very gentle fire, that gives not a stronger heat, than the Sun does on a hot day. Afterwards increase the fire by degrees, forcing at last the Spirits with the strongest fire, till red visible drops do come over, which work has taken up three days and nights. This being done, take that which is left in the distilling Vessel, commonly called *Caput mortuum*, and grind it small, pour on it clear Rain-water first distilled, and boil therein the Colcothar, and the Salt of the Vitriol will go into the Water. The Water being settled and clear, filter it, that the feces may be separated. Let the Water vapor away gently in some glass vessel, till the Salt be dry; dissolve the Salt again in Rain-water first distilled and let it vapor away again to dryness. Repeat this operation the third time, and the Salt of Vitriol will be very fair, clean, and clear. Put this dry Salt into a Cucurbit of Glass, and pour on it the above made Spirit of Vitriol, Lute the glass *lute sapientie*, and set it in digestion for some days. This being done, open the Glass, and put the materials together into a Retort of Glass, and distill them first gently, and when it ceases to drop, increase the fire, and force it over, till nothing will come more. Let it become cold, and then take the Spirit out of the Receiver, which must be somewhat large and strong. Put the Spirit into a Glass-body, and rectify it by distillation, till it be freed from the phlegm, and the matter in the glass-body appears to be of a red deep brown color. Then take the Glass-body, and set it with the said Matter in a Cellar, and there will shoot from it very fair, white, clear, transparent Crystals. Put these transparent Crystals into a large Phiol, with a very large and long neck, and pour on them the first white Spirit of Turpentine, and it will boil up and foam, therefore you must be very careful, and not over-hasty in doing this. The Crystals will dissolve, and the Spirit of Turpentine will grow transparent, as red as blood. This being done, pour on it three times the weight of common Spirit of Wine, freed fully from its phlegm, so that it stands two fingers high above it. Then put a little Head of Glass upon the Neck of the Phiol, luting it well, join to it a Receiver, and distill very gently the Spirit of Wine in *Balneo Mariæ*, and the tincture of Vitriol comes over very pleasant with the Spirit of Wine, and that which is corrosive remains behind with the oily parts of the Spirit of Turpentine. The Spirit of Wine being come with the Tincture, put it together into another Phiol, and pour on it some fresh Spirit of Wine, and distill it again gently in *Balneo Mariæ*, as you did before; if any corrosive be come over with the fire, it will now stay behind. Repeat this Operation the third time, and the work is done and perfect. Put this fair, red, transparent Spirit of Vitriol into a Pelican, add to it at once half an ounce of well-pulverized Unicorns-horn, and let it stand in circulation in a gentle heat for a whole month. Then pour it off very clear from the feces, and the Tincture of Vitriol is prepared for the Medicine, of a very pleasant taste, and is to be used after this manner following, to wit, let him that is troubled with the Falling-sickness, take half a dram of it in a spoonful of *Lillium convallium* Water, when the fit is coming upon him, thus let him use it three times, and this Medicine will cure him by the help of God. He that is mad and distracted, should take it likewise in Wine for the space of eight days, and he will have reason to give thanks to God for it. Moreover, if it be taken in Wine, it does resolve any hardness settled in the nerves, and if it be constantly used for some time, even the Gout itself is consumed and cured thereby.

Likewise it makes those who are melancholy and troubled with sadness, if it be used as before, very cheerful and lighthearted, dispels all sadness, and breeds good pure blood. It has been found very excellent in swimmings and giddiness in the head, it comforts the brain, and preserves the memory. If it be administered in Consumptions of the Lungs, arid any other Coughs, in the manner aforesaid, it will cure those distempers, and is very useful for many things.

An Addition.

Take Sal-armoniac, dissolve a considerable quantity of it in the strongest Vinegar, and add to it filings of Copper, let it putrefy in heat, till the filings are all grown friable, so that they may be ground into powder, and you will have a yellow powder, which Edulcorate well.

Having done so, dry the powder, and pour on it the red *Aqua vitae vitrioli*, which has been distilled over with its proper Salt, so that it covers it all over; set it thus in heat, and the powder of the Copper will be dissolved in the Oil, but there must be some fair Water mixed with it, then draw it off in sand to dryness, and the phlegm comes over: The remainder force out of a Retort in an open fire, and you will find an *oleum Veneris*, green, transparent, like an Emerald. Put again into this *oleum* some of the powder of Copper, and it will presently be dissolved in it. Then coagulate it to dryness, and you have a powder; half an ounce whereof will transmute a whole pound of Iron, being in flux, into very good Copper.

Of the Sweet Essence of Vitriol

The sweet Essence of Vitriol, whereby many wonderful cures may be wrought, is only prepared out of its Sulphur, which burns like other Brimstone. To obtain this, proceed after this manner. Take of the best Vitriol you can get; dissolve it in fair Fountain-water; after this take Pot-ashes, such as Dyers use for their Dying, those dissolve likewise in fair Fountain-water, let it settle well, and then pour off the clear from the dregs, and add to it the Solution of Vitriol, and one will inflame the other, and cause a separation. For the Sulphur of Vitriol does separate itself by precipitation. Make a considerable quantity of it, and Edulcorate it from all impurity. Afterwards dry the same Sulphur, which will burn like other Sulphur, being cast upon glowing coals.

Take now of this Sulphur, and sublime it by itself, without any addition, and there will remain some Feces, which separate and put away. Then take the Sulphur, and grind together with it half its weight of common Salt of Tartar, and distill them together through a Retort, and there will come over a reddish Oil. Pour to this Oil some distilled Vinegar, and there will precipitate a brown powder, and the Spirit of Tartar remains in the Water. Edulcorate the same powder very well, for therein is the treasure to be looked after. This work being done pour some Spirit of Wine on the said powder, and let it circulate in heat for eight days. Thus, the excellent sweet Essence of the Sulphur of Vitriol goes into the Spirit of Wine, and swims upon the top in *forma olei*, like Oil of Cinnamon. Then separate the Essence from the Spirit of Wine, by means of a separating Glass, and keep it very carefully for use, it being a great treasure.

The Use of this Medicine.

This Essence of Sulphur, four grains of it being taken in Balm-water, dries up the bad humors of the blood, strengthens and incites Men and Women to Copulation, cleans the womb, hinders the Rising of the Mother, and breeds good Seed for the Procreation of Children.

The same quantity being taken in Parsley-water, and continued for a fortnight, does consume all phlegmatic humors of the whole body; cures the Dropsie radically, drives out the putrefied Blood, opens Impostures, yea, you will find it really and in truth to does wonderful cures, if you will be industrious and careful in the preparation thereof, but you must never while you live, forget God your Creator, to call upon him for a blessing, and to render to him thanks for all his Fatherly benefits he has bestowed upon you.

Note, this sweet Essence of Vitriol has that Eminent Physician Doctor *Hartman* taken out of this book, and inserted in his *Praxis Chymiatrica, sub titulo lepra*, where he does explain something of this Process.

The Preparation of the Stone Ignis.

Now I will teach you the chiefest preparation of Antimony, and the use of it in Medicine. In this Antimony are hidden, and found so many wonderful mysteries, that there is none too old to learn, and to search to find them out. Therefore, will instruct you here to make only some preparations, which also are required to other things.

Take pure Mineral Antimony, which is brought from *Hungary*, grind it small, and wash it very clean, that the earth may be separated from it. Take then a pound of it, mix with it as much of fluxing powder made of Tartar and Nitre, cover it with common Salt, and melt it down in a Crucible with a strong fire: when it is well melted like Water, let it cool; put again to it the like quantity of new fluxing powder and melt it once again, and then the *Regulus* will be clear and pure. Add to this *Regulus* its weight of Nitre, and melt it down. Pour it out together, and beat off the *Scoriæ*, and put again to the *Regulus* its weight of Nitre, and melt it. Repeat this till all the *Regulus* is gone into *Scoriæ*, which you must carefully keep: They will burn upon the tongue like fire. This being done, take the matter so gathered grind it small, and edulcorate the Salt-peter from it, and there remains a brown-yellow powder, which dry and keep, it looks like ground glass. Take now a common *Regulus* of Antimony, made with Salt-peter and Tartar, grind it small, and put it into a round Glass, which must not be too high, and fasten a Head to it. Sublime your *Regulus* in sand by itself without any addition. Sweep the Sublimate with a Feather again into the glass, and Sublime it again; repeat this so long, till nothing do rise, but remain red and fixed in the bottom. Then take this fixed Antimony, and put it upon a Stone in a Cellar, and in time it will be dissolved into Water, which distill in *Balneo Mariæ*, until the sixth part only of the Water do remain in the Glass. Set this in a cold place, and there will shoot reddish Crystals, which dissolve in Rain-water, filter it, and draw off the Flegme to a thickness; set it by as before, and the Crystals will shoot white and very pure, like unto Salt-peter: This is the Salt of Antimony. Take these Crystals, and pour upon then pure distilled Vinegar, and they will dissolve in the Vinegar. Then distill the Vinegar, the Glass being very close luted, forcing at last the Spirits into the Vinegar, and then the Vinegar is prepared. Take this Vinegar, and pour it on the prepared brown-yellow powder, and set it in some warm place, and the Vinegar will draw out the Tincture of Antimony altogether red within half a quarter of an hour. Pour off this Extraction together, and set it to digest for eight and twenty days in

Balneo Mariæ. Afterwards distill from it the Vinegar through an Alembic in Sand, forcing in the end the Oil into another Glass, which comes over with many strange and wonderful Veins. Rectify this Oil in Ashes, and the rest of the Vinegar, if any be left, will come off, and the Oil remains very sweet, and of a pleasant red color like a Ruby. Thus, have you joined the Sulphur with the Salt of Antimony, and brought it over like an *Aqua vitæ*, which keep very carefully.

Furthermore, take again a common *Regulus* of Antimony, made with Salt-peter and Tartar, and beat it to powder. Then take of strong distilled Vinegar three measures, [*alias*, one measure] *id est,* four quarts and a half. Put into it of Sal-armoniac: Of Salt of Tartar (wherewith I will teach you hereafter at the end of my directions where I intend to write of the Philosophers Stone, to prepare Spirit of Wine) likewise eight ounces. Digest this to the Evaporation of the Vinegar, and mingle with the Salts three parts of *Venice Tripoly*, and distill the Spirit, which is of a singular nature and property. Pour this Spirit on the pulverized *Regulus* of Antimony, and having the Glass well Luted, let it stand in digestion sixteen days; then distill the Spirit from the Matter to a dryness, and grind four times the weight of Filings of Steel with the same, put it into a Retort, and putting thereto a large Receiver full of Water, distill it, forcing at last with a strong fire, and the Mercury comes over in Fumes, and is quickened in the Water, which is the true Mercury of Antimony.

Take common Spirit of Vitriol, add a little common Water to it, and put your Filings of Steel into it, let it stand till the Filings are dissolved, then pour it off clean, and put away the Feces. Afterwards distill the Spirit in ashes to a thickness, and set the Glass in a cold place, and there will shoot good Vitriol of Iron, which take, and having first vapored away the phlegm, mingle with it three parts of the powder made of burned Potshards of broken Pots: put it into a Retort, draw off the Flegme first, then force the Spirit with a strong fire into a proper glass, which rectify to the height, and there will remain an Oil in the bottom. Pour this Oil upon the Mercury made before, and draw off the Flegme in hot ashes, and the tincture of the *Aqua vitæ* remains behind, and does precipitate the Mercury into a fair high colored powder of very great virtue in curing old Running Sores.

The Conjunction of the three Principles, Sulphur, Salt, and Mercury of Antimony.

Take then of this Precipitate well Edulcorated with common Spirit of Wine, one part; and pour on it of the above mentioned sweet Oil, three parts, in a Phiol, so that the Phiol be not above half full.

Then seal it Hermetically, and place it in a Philosophical Furnace, and the precipitate will be dissolved in that continual heat. Open then the Glass, and continue a strong fire, till the Matter becomes a fixed Powder, and do fix, and then the Stone *ignis* is prepared, of which I have written. This Stone is a peculiar Tincture in Men's bodies as well as in those of Metal. This may be used in many hard and dangerous distempers, as I have set down in the directions for the administration and use of the same in *The Triumphal Chariot of Antimony*.[23]

An Addition.

Take of this Stone, or particular Tincture, half an ounce, cast it upon twelve ounces and a half of pure Silver, or upon as much Pewter or Lead, let it flow very well for four and twenty hours; then drive it off clean, and Quart it, as Tryers and Refiners do, and you will find in the Silver two ounces and a half of very good Gold, and in the Pewter or lead one ounce, upon the Cupel.

Another Medicine made of Antimony and Mercury
And of its Effects in Outward Sores.

Take *Hungarian* Antimony, and Sublimed Mercury, and grind them well together, and distill them through an Earthen Retort, forcing them at last with the strongest fire imaginable, and you will obtain an Oil, which separate and keep apart. Put away the quick Mercury, if there is any, and the Cinnabre you will find in the neck of the retort. But as for the *Caput mortuum*, grind it small, and put it into a new retort, and having poured on it the Oil, first made warm, distill it again from it. Repeat this so often, till the *Caput mortuum* remains behind like Ashes, and then your Oil is prepared. After this take so much fresh Antimony, as first of all the *Caput mortuum* did weigh, grind it small, and pour on it the Oil first warmed, and so many times distilled as before, till the Oil comes over as red as a Ruby, and the *Caput mortuum* likewise remains like ashes in the bottom of the Glass, and then the Oil is prepared.

The Preparation of the Sublimate for this Work

Take one pound and a half of *Hungarian* Vitriol, one pound. of common Salt, four ounces of Salt-peter, grind this together, and put one pound of Quick Silver into the bottom of a Glass body; place it in Sand, so that the Sand does not come above the matter in the Glass; put a Head thereupon, and give it a convenient fire,

---

[23] Volume 2 in The R.A.M.S. Library of Alchemy.

and the sublimate will stick to the sides of the Glass, which is to be used to your work.

Take the above prepared *Aqua vitae*, and add to eight ounces of it, three ounces of Salt-peter-water, and distill it out of a Coated Glass Retort, and you will have an ounce of the *Aqua vitae* remain behind fixed. Then put again to the *Aqua vitae*, one ounce of fresh Salt-peter-water into a Retort, and distill as before, and there will stay more behind. This addition of fresh Saltpeter-water to the *Aqua vitae*, as distillation out of a Coated Retort, as has been said before, repeat so often, till all remain fixed in the Retort.

The Salt-peter-water is made thus.

Take unburnt Potshards ground small, and with three parts of the same, grind one part of purified Salt-peter; put into the Receiver half a pound of Water to one pound of Salt-peter, and force the Spirits over into it. That which is fixed with this Water, put into a Glass body, and pour upon it the common *Aqua vitae vitrioli*, so that it be four fingers high upon it. Then distill it till the Matter becomes dry. Take out this matter, and dry it yet more, that the rest of those corrosive Spirits may evaporate; then edulcorate it well with Spirit of Wine, and the Medicine is prepared.

The Use.

Three or four grains of this Medicine being taken in some good Treacle for some days, cures the French-pox, nay, there is no sore so old and festered, but is cured infallibly by it. I have cured with it likewise many spreading old running Ulcers, as Fistulas, Cancers, the Wolf, and the like: For which many have with their prayers given thanks to God, and me for the Physick. The name of the Lord be praised therefore. AMEN.

Note, this Essence of Antimony has been prepared by divers Eminent Physicians in *London*, which had communicated to them this process out of the *High-Dutch*, and has been used with very great benefit in many desperate distempers, and diseases which were accounted incurable.

Though many more, yea, numberless Medicines, may be prepared out of Antimony, as *Aqua vitae*, Powder, Extractions, Vitra, and the like, of which you see my **Triumphal Chariot**: Yet have I set down here only such, as will be a sure remedy in any distemper almost, as well inwardly as outwardly applied.

The Preparation of a Medicine out of Common Sulphur

Take common Sulphur, and grind it small. Then grind with it three parts of calcined Vitriol, put it together into a high Cucurbit, and Sublime it in Sand, till nothing will Sublime more. Take then these Flowers, put them into a Glass, and pour on them a common *Aqua vitae Tartari*, which has been dissolved in a Cellar, so that it swims on the top of it a hands breadth. Place it in a convenient heat, and the Sulphur will open itself in few hours, and become transparent red like a Ruby. This being done, pour off the extraction into another glass, and put to it very good distilled Vinegar, and the Sulphur falls to the bottom with a great stink. Pour off the *Aqua vitae*, and edulcorate well the Sulphur, and dry it gently. Put this Sulphur again into another Glass Cucurbit, and pour upon it Spirit of Wine, which is prepared with Philosophical Tartar; set it in heat for three days, and the Spirit of Wine imbibes again that excellent Tincture of the Sulphur: Then pour off the Extraction, and draw off the Spirit of Wine with a pretty strong fire in Sand, and there will come over with it a pleasant sweet smelling *Aqua vitae*. Having done so, rectify the Oil in *Balneo Mariæ*, and draw off the Spirit of wine gently, and the *Aqua vitae Sulphureous* remains in the bottom.

The Use of this Medicine.

Six or eight drops of this Oil being taken in a Spoonful of Wine, are good for those that are in a Consumption: It is good likewise for Coughs, opens the Breast, and Ulcers of the Breast, and also Imposthumes; it relieves against whatsoever may occasion any putrefaction in a man's body, if use of it be continued for some time.

The Preparation of the Tincture of Corals.

Take red Corals, break them into pieces, and pour on them a common Spirit of Salt, and the Corals will be dissolved. This being done, draw off by distillation the Spirit of Salt, and Edulcorate them well. Then take to one Marck of this powder, half an ounce of common Sulphur pulverized, and having mingled it together, reverberate it very gently, till all the Sulphur be burnt away. Having done so, grind as much Camphire with the Corals, and burn the Camphire likewise away. Then Edulcorate well the Corals, and pour upon them high rectified Spirit of Wine, and digest them for eight days, and the Tincture of the Corals will Elevate itself, and go into the Spirit of Wine. Then pour off that which you have extracted, and after that draw off the Spirit of Wine from it, and there remains the Tincture of Corals behind in the bottom like a red fat Oil of Olives.

The Use of the Medicine.

Six drops of this Tincture given in a Spoonful of Wine to those that are bereaved of their Senses, restores them again. This Tincture comforts likewise the Brain, and strengthens the Memory, dispels sadness and melancholy, makes lighthearted, breeds good blood, and strengthens the heart. It is such a noble Medicine, for which we are bound indeed to bless Almighty God.

Of the true Solution of Pearls.

Take very good Verdigrease, grind it small, and dissolve it in distilled Vinegar, pour off the clear, and throw away the Feces. Then distill off the Vinegar out of a Glass body to a thickness, and put it into a cold place, and there will shoot from it a fair Vitriol: Put this Vitriol into another glass, and pour on it a highly-rectified Spirit of Wine, and dissolve therein the Vitriol very well; separate the Feces from it, afterwards distill off likewise the Spirit of Wine to a thickness, and set it again into a cold place, and the Vitriol shoots again. Put then the Vitriol into a Glass body, and draw off by distillation the Flegme in *Balneo Mariæ*, till the Matter becomes dry; take it out, put it into a glass Retort, and distill once more with a stronger fire in sand, and you will obtain a pleasant Vinegar. Dissolve in this Vinegar as many Pearls as it will dissolve, for this Vinegar works very well upon them, dissolving the substance, but not the shells.

The Pearls being dissolved, draw off the Vinegar in *Balneo Mariæ*, till the Pearls are very dry: Then take them out, and edulcorate them with Rose-water. Put these Pearls thus prepared into a Glass body, and pour some Spirit of Wine upon them, and digest them in gentle heat four and twenty hours, and there rises a pleasant liquor from the Pearls, which does mount and swim upon the Spirit of Wine like an *Aqua vitae* made of Cinnamon. Pour it off together with the Spirit of Wine, and keep it.

The Use of this Medicine.

Take of this Spirit of Wine half a spoonful, so that four or five drops of the Oil may go with it: It comforts the Heart, gives strength to the very Marrow and Bones; cures Swimmings in the Eyes, and whatsoever may be hurtful to the Eyes.

Dispels Rheums in the Head, and the Noise in the Ears, opens the passage to Hearing, and is moreover a most precious treasure in many distempers.

Note, this preparation of Pearls has been borrowed of our Philosopher, by that Illustrious Reformer of *Galenical* Medicines Doctor *Zwolffer* in his Appendix.

A certain Cure of the Stone.

Rx. of common Salt-peter well purified one pound, and as much of the common white Spirit of Vitriol. Pour the Spirit of Vitriol upon the Salt-peter, and the Salt-peter will be dissolved altogether. This being done, distill from thence the Spirit of Vitriol in ashes, to a thickness, and set it into some cold place, and the Salt-peter will shoot again from it. Take two ounces of this Salt-peter, and the like quantity of the Salt of Wormwood; pour on them a little of the Oil of Sulphur made *per Campanam*, so that the Salts may be like a Pultise: Mix with it likewise one dram of Aniseed-Oil, and as much of Oil of white Ambre, adding thereto a pound of Canary Sugar, and mix all these ingredients very well together. Let him that is tormented with the Stone, take of this Powder every day five or six times, every time as much as will lie upon the point of a knife, twice repeated, and this Medicine will work upon the Stone and break it, and throw it out radically. I have done great cures with this Medicine, for many have been cured by it. Yea, in the beginning of my practice, I have cured one of my Brethren of his distemper with the said Medicine, when all the Herbs he used would do him no good. He prayed fervently for me to his dying day, and gave God thanks daily for his Creatures, seeing he had put so great virtue into them.

Note: With this Medicine, very great cures have been performed, as concerning the Stone of the Kidneys, by divers excellent Physicians in *Germany*.

Of the Soul, or of the Sulphur of Lune
Or the Philosophers Silver.

Take common Salt and quick or unslaked Lime, reverberate them together in a Wind-furnace with the strongest fire, extract again the Salt-peter with warm Rain-water, and coagulate it to dryness, mingle again with it new quick Lime, reverberate it, and extract again, repeat this the third time. This being done, take Calx of Silver, being after the dissolution in *Aqua fort* precipitated, and mix it with the prepared Salt: Put it into a glass Phiol, pour on it a common *Aqua fort* such as the Goldsmiths use, made of Salt-peter and Vitriol, and draw it off by distillation in hot Sand, pour on it some fresh *Aqua fort*, and having distilled it likewise, repeat it the third time, giving at last a very strong fire, that the Matter in the Glass may flow very well. Let it cool of itself in the Furnace, and the Silver will become transparent blue in one piece. Extract this with Vinegar, till you can extract no more. Edulcorate that which is extracted with water, that the Salt may be separated from it. Cohobate Vinegar upon the dry Sulphur, till it comes over like a Saphire. Reduce the same Silver into small filings, and add to its weight of Sal-armoniac, and Sublime it in a Glass body, and the Sal-armoniac carries with it the Sulphur of Lune, of a very pleasant Sky color. Put this Sublimate into a Dish of Glass, Edulcorate it well with Rain-water first distilled, and the Sal-armoniac will be separated. Then dry the Sulphur of Lune, put it into a little body, and pour on it good rectified Spirit of Wine, and set it four and twenty hours in heat, and the Spirit of Wine does Imbibe the Sulphur of Lune fine transparent blue like a Saphire, or *Ultramarin*, and leaves some few Feces behind, which separate from it.

The Use.

Five or six drops of this Tincture being taken in Wine, do dispel sad and melancholy thoughts. It prevents unquiet sleep; cures those as use to rise and wander up and down in the night, and likewise those that are Lunatics. Gives rest to all such as are restless in the night, and is an excellent Medicine for all those that are Lunatics.

The Secret of Quick or Unslaked Lime.

Take good pure Chalk, burn it in a Potters Furnace with a very strong fire, to bring it to an exact maturity. Then grind it small upon a warm Stone, and pour on it in a Glass body Spirit of Wine, made with Philosophical Tartar, as I shall teach in my way of making the Potable Gold, that the Chalk become like a thin Pultice.

This being done, distill from thence the Phlegm, to the dryness of the Chalk, pour fresh Spirit of Wine on it, and distill it off again. Repeat this six times; then grind the Matter small, and lay it on a Stone in a Cellar to dissolve, and there will flow in a few days from it a Liquor, which when you have gathered, put it into a Retort of Glass, and distill it in Sand; and the phlegm comes over first, which keep apart. After this there comes a spiritual liquor, which is likewise to be kept by itself. Moreover, take Crystal-stones, pulverize them, and grind their weight of Live or Mineral Sulphur with them. Put then this matter upon a broad earthen Platter, stirring it continually, and burn away the Sulphur from it. Then Reverberate it in an open flaming Fire for three hours. This being done likewise, put the Matter into a Glass, and pour the liquor upon it. Take likewise Crabs Eyes, put them into another Glass, and pour on them of the same liquor: Let it stand pretty hot for fourteen days and nights, and there will rise from both a moisture, which pour off together very clean into a little body of Glass, and rectify it in *Balneo Mariæ*, and the Liquor remains behind. Three grains of which being taken in wine, has wrought very great and admirable effects.

This Medicine cures likewise radically the Stone of the Bladder and Kidneys both in Men and Women.

An Addition.

Take this burned Chalk, pour upon it, and then draw from it again several times an *Aqua fort* made of Vitriol and Salt-peter. Dissolve it afterwards in a Cellar. Distill that which is dissolved into an Oil with a strong fire. Digest with this Oil a Calx of Lune opened with *Aqua fort* for a Month. Reduce this Calx by melting it down with Salt-peter, and Sal armoniac, and refine it with Saturn, then separate it, and you will have a white fixed Lune, which lay for a day and night in an *Aqua fort*, and you have good Gold which endures all trials. L. D.

The Preparation of the Great Philosophic Stone.

Lastly, to close all, I will now instruct you from the love I bear to God, how I have made my Universal Medicine, or the Philosophers Stone, which many Master-builders have wrought upon, and will discover faithfully and truly all my Manual Operations. You must know, that our Stone is made out of its own proper Essence, for it transmutes other Metals into real and true Gold, which Gold must be prepared, and become a better Stone. And though nothing of another Nature must be used in the preparation of our Stone, which might obstruct its Majestic Excellency, yet the preparation of it in the beginning cannot be made without means. But observe, that, as you will hear afterwards, all corrosives must be washed away again from it and separated, so that our Stone may be severed from all poison, and be prepared to be the greatest Medicine.

But I beseech you for God's sake, that you will keep your Tongue, and put a seal to your Lips, that you may not discover what you learn out of this, such an incredible worldly excellency, to the impenitent and unworthy, that you may not participate of other Men's sins where there is no need, and thereby prepare for yourself the way to Hell and everlasting Damnation, which God Almighty graciously keep and preserve you from. Wherefore observe my words, and lay to heart my sayings, do not dislike it, that I simply relate these things, for many words avail little here. Neither let it trouble you, that the work may seem slight unto you, but consider the end that will follow. For inconsiderable is both the beginning of the work, and the work itself: but the end is high and excellent, all which knowledge and experience will discover and bring to light.

Now do I proceed in the Name of the Lord to the Work itself.

Take of the very best Gold you can have one part, of good *Hungarian* Antimony six parts, melt this together upon a fire, and pour it out into such a pot as the Goldsmiths use; when you have poured it out it becomes a *Regulus*. This same *Regulus* must be melted again, that the Antimony may be separated from it.

This being done, add to it Mercury, and melt it again, and cleanse it again. Repeat this the third time; and the Gold is purged and purified enough for the beginning of the Work. Then beat the Gold very thin, as Goldsmiths do, when they gild, and make an *Amalgama* with common Quick-silver, which must be squeezed through a Leather; let the Quick-silver fume away by little and little upon a gentle fire, that nothing of it may remain with the Gold, and stir it about continually with a small Iron, and the Gold is become subtle, so that its Water may the better work upon it, and open it.

The Preparation of the Water.

Take one part of Salt-peter well purified, and grind with it the like quantity of Sal-armoniac, and half as much of Pebbles very well cleansed and washed. Mingle all these ingredients together, and put them into an Earthen Retort, and the Spirits may not come through, put the same into a Distilling Furnace: The Retort must have a Pipe behind, and put as large a Receiver as you can get to the Retort. The Receiver must lie in a Vessel full of cold Water, and a wet Linen-cloth must be put round about it, which you must wet continually with another wet Cloth; then put again so much Matter into the Retort, till all is gone into it, and then your Water is prepared.

Take then of the prepared Calx of Gold one part, put it into a Glass body, and pour three parts of the above made Water upon it, and place it in warm Ashes, and the Gold will dissolve in it; but if it should not altogether be dissolved, pour more fresh Water upon it, and it will dissolve all. This being done, pour it out into another Glass, and let it stand till it becomes cold, and it will let fall some Feces, which separate by pouring the Water from them into another Glass, set this Glass in *Balneo Mariæ*, and put a Head upon it, let it stand in heat day and night, and more Feces will settle, which separate from it as before. Close-up your glass very well after you have put on the Head, and Lute another Glass to the Head, and let it stand for fourteen days in a gentle heat, that the Body may well be opened. This being done increase the fire, and distill off the Flegme to a thickness, that it remains in the bottom like an *Aqua vitae*. That which has been distilled, pour again into the body, having first made it warm, and Lute again the Head to it, and let it stand to digest a day and night. Then draw off the Water again by distillation, and pour it again warm upon it. Repeat this so long till the Gold is come over altogether into a low body with a flat bottom. Put this spiritualized Solution of Gold again into a Glass, and pour on it a considerable quantity of Rain-water, putting thereto three parts of live Mercury to one of Gold: But you must squeeze first the Mercury through a Leather, and stir it very well together, and you will see many wonderful colors, and if you do repeat this, stirring several times, there will fall an *Amalgama* to the bottom, and the Water will become clear.

This being done, decant the water, and dry gently the *Amalgama*, which having Edulcorated very well, put it upon a broad shallow Earthen Platter, under a Cover, stir it about continually with an Iron Wire, till all the Quick-silver be fumed away, and there will remain upon the Earthen Platter a very fair Powder of a purple color.

Afterwards you must prepare your Spirit of Wine
With the Philosophical Tartar,
In the manner following.

First of all, you are to know, that the Tartar of the Philosophers, whereby the Lock is unlocked, is not like unto common Tartar, as many do think; but it is another Salt, and springs from one root; and this is the only Key to open and to dissolve Metals, and is prepared as follows. Take Ashes of a Vine, which has born Grapes, that have yielded good Wine; make of them with warm Water as strong a Lee, as possibly can be made. When you have a considerable quantity of this Lee, boil it away, and coagulate it to a dryness, and there remains a reddish Matter. Put this Matter into a Reverberating Furnace, and reverberate it for three days, or thereabouts, in an open fire, that the flame may play very well upon it, and stir it continually, till the Matter is become white. Afterwards dissolve this reverberated Matter in Fountain-water, and let it settle, pour off the clear, and filter it, that all the feces may be separated, and coagulate it in a Glass-body, and you will have a pure white Salt of Tartar, from which a true Spirit is drawn.

Take now high rectified Spirit of Wine, fully freed from its phlegm, put the same into a Glass-phiol, with as long a Neck as possibly you can get. But first of all put into it your Salt of Tartar, and then the Spirit to the supereminency of three fingers; Lute a Head to the Phiol, and put thereto another Glass, let it stand in a gentle heat, then distill gently off the phlegm, and the Spirit of Tartar is opened by the Spirit of Wine, and by reason of their reciprocal wonderful love, it comes over with the Spirit of Wine, and is united with it. The remaining feces, and some phlegm staying behind with them, are to be put away.

This is now the right Spirit of Wine, wherewith you may open that which the Lover of Art desires to know, for it is become penetrant by preparation.

Take now the powder of Gold of a purple color, and having put it into another Phiol, pour on it your Spirit of Wine; put it very close Luted in a gentle heat, and it will extract the Sulphur of Gold within four and twenty hours, of a high red color like blood. Having done so, that it does not yield any Tincture more, pour off the extraction very clear into a little Glass body. The remainder is a white Calx, pour upon this Calx the afore said Spirit of Wine, and let it stand in putrefaction, having the Glass well-stopped for fourteen days and nights; and the Spirit of Wine will become of a white color like milk, which pour off clear, and pour upon it fresh Spirit of Wine, let it stand a day and night longer, and it will be colored again, but not so much; add this to the first, and what remains do not dry, but leave it in the Glass. Put the white Extraction into a little body, and distill the phlegm from it, till it be reduced to a small quantity.

This being done, put the Glass in a Cellar, and there will shoot from it fair and transparent Crystals, which having taken out, put the remainder again in a Cellar, and you will have more Crystals, which put together into a body of glass, for it is the Salt of the Philosophers, and pour half the extraction of the Sulphur of Gold upon them, and they will dissolve immediately, and melt like butter in hot water: And then distill it together out of a Glass-body in hot Ashes, and it will come over together in a form of a red Oil, which falls to the bottom, and the Spirit of Wine swims upon the top which separate from it.

This is the true Potable Gold, not reducible into a body, and my *Phalaja*, whereby I have cured innumerable People, administering but three grains of it in Wine.

The other half of the Extraction must be distilled gently in *Balneo Mariæ* to a dryness, that the Spirit of Wine may be separated. Pour on it this Oil of Gold, or Potable Gold, and it takes up the powder in a moment, and becomes a much higher color than it was before; and this will dissolve in common Spirit of Wine, and other Wine, as red as a Ruby; which constantly and wonderfully cures all such distempers of the body, as have their original from within.

Then take that other part of Mercury of pure Gold, which you have kept, and pour all this being its own Oil upon it; and distill by an Alembic, but not too strongly, and there comes over some Phlegm, and the Oil does precipitate its own Mercury, and becomes white again, the greenness being lost and gone.

This work being done likewise, get a Philosophical Egg, which the Philosophers call their Heaven, and you will find two parts of the Oil in weight to one part of the precipitated Mercury. Put then the Mercury into a Glass, and add the Oil of Gold to it, so that one part of the Glass may be filled, and three parts remain empty. Seal it well, as *Hermes* teaches, and put it into the threefold Furnace, so that it stands not hotter than an Egg, which is under a Hen to be Hatched; and the Matter will begin to putrefy within a month, and become very black; which when it does appear, it is then certain that the Matter is open by the putrefaction, and you may be glad of that happy beginning. Increase now the fire to the second degree, and the blackness will vanish away in time, and change into many admirable colors. These colors being gone likewise, increase the fire to the third degree, and your Glass will look like Silver, and the Rays will become ponderous. Then increase the fire to the fourth degree, the Fumes will cease by little and little, and your Glass will shine as it were beset within with a Cloth of Gold. Continue this fire, and the Rays will disappear likewise, and there will be no more Rays be seen to rise, but you will see your Matter lie beneath like a brown Oil, which at length, being become dry, does appear like unto a Granet, which is both fixed and liquid like Wax, penetrant like Oil, and mightily ponderous.

He that has obtained this, may render thanks to God his Creator, for poverty has forsaken him: Diseases will fly from him, and wisdom has taken possession of him. For this noble Medicine is such a Stone, to which nothing in the world may be compared for virtue, riches, power, honor and might; but it is to be preferred before all earthly things, which the whole Universe does comprehend.

Having thus prepared your Medicine, if you intend to multiply it, you must proceed as follows. Take of the prepared powder of Gold, of a purple color, as you have done before, three parts. Add to it of the prepared Tincture one part, in a new Heaven, or Philosophers Egg; seal it again Hermetically, and set it into the Furnace, as before, and the Matter will unite itself and dissolve, and be brought to perfection within thirty-one days, which is a month, which otherwise will take up ten months. Thus, you may multiply your Medicine *in infinitum*, so that you may perform things, which the world will account incredible.

The most High God and Creator be blessed and praised for evermore for his inexpressible benefits, and likewise for all his gifts and wisdom he has been pleased to bestow.

Lastly, you must know, that this Medicine is a very spiritual and piercing one, which cures any distempers of this world, in all Creatures whatsoever they be. One only grain of it being taken, it penetrates the whole body like a fume, chases out of the body all that is bad, and brings that that is good in the room of it renewing the man, and makes of him as it were a new man, which it preserves without any accidents to his age, and the teem prefixed by the most High. *Contramortem enim remedium non est.*

This Medicine being first fermented with other pure Gold, does likewise tinge many thousand parts of all other Metals into very good Gold, as the last Key of mine, which is the twelfth, teaches by a certain way and process, whereby such Gold likewise becomes such a penetrant Medicine, that one part of it does tinge and transmute a thousand parts of other Metals, and much more beyond belief, into perfect Gold. God be blessed and praised both now and forever more. *Amen.*

An Addition
Of
Basil Valentine,
Which he himself has annexed to his
Manual Operations,
For a fuller declaration of the same.

To make a true Spirit of Wine.

Take *Vinum Adustum*, and put it into a strong Vessel, which endure the fire; light it with a Match of Brimstone, and put quickly a Head of Iron or Copper upon it; and the true fiery Spirit will be resolved into a Water in the Receiver, which must be large.

This is the true aerial fiery Spirit of Wine.

Spirit of Mercury by itself, or Mercurial Water.

Put running Mercury into a Retort, and put to it a Receiver, which must stand in a Glass with Water in it. Distill then, and the Spirit will precipitate itself, and is resolved into a Water. Pour out this Water, and put the Mercury, which sticks to the Neck of the Retort, back again into the Retort. Distill and rectify, till you have brought and reduced it to a water. This Spirit of Mercury cures almost all distempers, and does extract the essence out of Minerals and Metals.

A Tincture both upon Men and Metals.

Take the spiritual Gold of a purple color, extract its Sulphur with distilled Vinegar, separate the Vinegar again from it, that it becomes a Powder. This Powder being dissolved in Spirit of red Mercury, that is, Gold, put thereto Salt of Gold, and fix it. This is a Universal Medicine for sick and diseased bodies of Men, and likewise to exalt Metals to the highest degree.

A Tincture upon White.

Calcine Silver with Salt and Quick-lime, and extract its blue Sulphur, which elevate and rectify with Spirit of Wine, that it remains a Liquor. Dissolve this in the white Spirit of Vitriol, and in the Spirit of Mineral Mercury.

I do not understand here the red Mercury, but the common white Mineral Mercury, or rather that is extracted out of Vitriol.

Fix it then, and you have a Universal Medicine against all distempers, and a Tincture, which does tinge Lead, Pewter, Mercury, and Copper into Silver.

To make an Ounce of Gold out of half an Ounce.

Take Spirit of Salt, rectify it with Spirit of Wine, that it become sweetish. Pour this upon the Spiritual Gold of a purple color, and it will extract only the Soul or Sulphur of the Gold, but does not touch the body of Gold. The Sulphur of Gold does graduate Silver into Gold yet no greater quantity of it, then there has been in the Gold. The Body of Gold must be as white as Silver. Reduce it upon a Cupel with Saturn, and a little Copper, and the white Body of Gold does recover again its color and property, and becomes good Gold.

To make the Mercury of Gold, or the Philosophical Mercury

Take the Gold of a purple color, out of which the Sulphur is already extracted, digest it with the following Water for a month, then revive it again by driving it through a Retort, in the Neck whereof are to be laid thin Iron-plates, drive it into a Receiver with some Water in it, and it runs together, and becomes a quick Mercury of Gold.

The Water is made as Follows.

Take Salt of Urine of a Young Man, that drinks nothing but Wine; and likewise Salt of Tartar, and Sal-armoniac, *ana*. Let all this dissolve into a liquor, which rectify with Spirit of Wine, that it becomes very sweet.

This is the Arcanum, wherewith the body of Gold is reduced into running Mercury.

To make the Salt of Gold.

Pour Gold three times through Antimony, beat it into thin Plates, and dissolve them in *Aqua Regis*. Dissolve likewise Salt of Tartar in Spirit of Wine, and draw off the phlegm, that it remains like an Oil. With this Oil precipitate your Gold, and separate again the Salt of Tartar from it by ablutions. Then reverberate it fourteen days. Pour upon this Calx of Gold, distilled Vinegar, let it boil gently a day and night, and the Vinegar does dissolve the Salt of the Reverberated Gold. What remains in the bottom undissolved, must be reverberated again for eight days. Then boil it again in new Vinegar, put this afterwards to the first solution. If anything remains yet behind, it must be reverberated eight days more, till the body is gone into the Vinegar. Then draw off the Vinegar in *Balneo Mariæ*, and you have the Salt of Gold in a yellowish powder, which cures all distempers.

The Second Treatise

Of Things Natural & Supernatural.

As also

Of the First Tincture, Root, and
Spirit of Metals and Minerals; how
They are Conceived, Ripened, Brought-
Forth, Changed, and Augmented.

Faithfully
Discovered by Basil Valentine Fryer of the
Order of the Benedicts.

Chapter I. Of Things Natural and Supernatural.

Because I purpose to write of the First Tincture, and Root of Metals and Minerals, and to give an account of their spiritual Essence, how the Metals and Minerals are in the beginning spiritually conceived, and corporally brought forth: It is therefore necessary to premise something, and to inform you in few words, that all things are divided into Natural and Supernatural, and whatsoever is Visible, Comprehensible, and Formal, the same is Natural: But that which is Incomprehensible and Spiritual, the same is Supernatural, and must be apprehended and judged of by Faith; as there is the Creation, and first of all the Eternity of God, which is Infinite, Unsearchable, and Immeasurable, which Nature is not capable of, neither can Human reason comprehend it. This now is Supernatural, which transcends reason, and is apprehended by Faith, which therefore is a Divine thing, and belongs to Divinity, which judges of Men's souls.

Secondly, and moreover there belong to Supernatural things the Angels of God, which have clarified bodies, and do perform such things by permission of their Creator, which otherwise no Creature is able to do, because their works are hidden from the eyes of all the world; as likewise the works of the hellish Spirits and Devils are abstruse, which they work by the providence of the most high God. But above all the great works of God are found and discovered to be Supernatural, because they cannot be comprehended, nor judged of by the thoughts of Men; as there is especially the mercy of God, and his great benefits, whi.ch out of his tender Love he shows unto, and bestows upon Mankind, which indeed no Man is capable to discern and know comprehensively: And likewise other great Miracles wrought at sundry times by Christ our Savior and Redeemer, to the confirmation of his Omnipotency and Majesty; As where he had raised *Lazarus* from the dead; and likewise the Daughter of *Jairus,* Ruler of the Synagogue, and then the Son of the Widow of *Nain.* He has made the Dumb and Speechless to Speak, the Deaf to Hear, and the Blind to See; which indeed are all Supernatural things, and wonderful works of God: As also his Conception, Resurrection, Descension to Hell, and Ascension to Heaven; which are too high and abstruse for Nature, and are to be attained unto only by faith.

Amongst Supernatural things are likewise to be counted *Elius,* and *Enoch's* Ascension into Heaven. The Rapture of Saint *Paul,* who was taken up into the Third Heaven in the Spirit. Furthermore, there are many Supernatural things performed by Imagination, Dreams, and Visions, as very often many Miracles are wrought by Imagination, as the speckled and spotted Sheep were conceived by the speckled Sticks laid in the Water. The wise Men from the East were warned by God in a dream, not to return again to *Herod*. Their three Persons likewise, and their three Gifts and Presents have a peculiar and mystical sense; as also the Supernatural Star. And the Dream of *Pilat's* Wife, who falsely condemned Jesus Christ our Lord and Savior, has not been Natural. Neither can the Vision of the Holy Angels, who appeared to the Shepherds at the Nativity of Christ, and to the Woman, who came to the Sepulcher of Christ, to see where his Body was laid, be accounted Natural.

Moreover, there have been often wrought many Supernatural things by the Prophesies of the Saints and Prophets. Thus was the voice of the Ass that spoke to *Bileam*, not agreeable to common Nature; and interpretation of Dreams by *Joseph* was likewise Supernatural. God preserves us many times by his holy Angels, from great and sad accidents, and delivers us from perils and dangers, which otherwise naturally would be impossible.

All this, and what else there may be, does belong to Divinity and Heavenly things, and are to be regarded by the soul. There are besides these other Supernatural things of the visible Creatures of God, as we do find, see, and discern in the Firmament, Planets, and Stars, together with the Elements, which indeed are beyond our reason, their course only being observed by the speculations of Arithmetic. This now belongs to Astronomy, and is a visible, nevertheless incomprehensible being, performing its operation magnetically, wherein likewise many miracles are discovered and found, which are altogether Supernatural. For you are to understand, that Heaven works upon Earth, and the Earth keeps correspondences with Heaven: For the Earth has likewise seven Planets in it, which are brought forth and wrought upon by the seven Heavenly Planets, only by a spiritual impression and infusion; and in this manner, all the Minerals are wrought by the Stars. This now is done spiritually beyond our apprehension, and therefore to be accounted not Natural, in the manner of two, that are enamored. The Men are visible, the Love is invisible, which they bear one to another. The Body of Man is Spiritual and Natural; but the Love is Invisible, Spiritual, Incomprehensible, and Supernatural, comparable to nothing else than to a Magnetic attraction. For the invisible Love, which out of affection is spiritually attracted by imagination, is perfected by consummation.

In like manner, when Heaven bears love to the Earth, and the Earth has love, inclination, and affection for man, as the great World for the little one, because the little World is taken out of the great, and when the Earth through the desire of an invisible imagination does attract such Love of the Heavens, then is there a conjunction made of the Superior with the Inferior, like unto a Husband and his Wife, which are accounted one body: And after such a conjunction the Earth becomes impregnated by such infusion of the Superior Heaven, and begins to bear a birth, according to the infusion, which birth is ripened, after its conception, by the Elements, and is digested to a perfect maturity. This is likewise numbered and accounted amongst Supernatural things, *viz.* how the Supernatural essence does perform its operation upon the Natural.

There are furthermore numbered amongst Supernatural things, all Magical and Cabalistic things, being subject thereunto, which do spring from the light of true knowledge; not such as have their original from Superstition, or unlawful conjuration used by those that conjure the Devil. This Magick is here not understood by Me; but such as was practiced by those wise men or Magi, which came from the East, who gave their judgement of things by the inspiration of God, according to the true and unforbidden Art, which likewise was used by the Ancients, the *Egyptians* and *Arabians*, who before writing was invented, did note, observe, and remember their things by Signs and Characters.

Neither is the use of such Blessings forbidden, which Christ the Son of God himself did make use of, as the Scripture does tell us: And he took the Children, laid his hands upon them, and blessed them. But those that are against God, and his word, are justly to be rejected, and not to be permitted, for they are not Divine, but Devilish; but such Supernatural things, which are not contrary to God, nor his Holy Word, they belong to Magick, which is no way hurtful to the Soul.

Concerning Visions, Apparitions, and the like, for they have often happened unto Holy Men, they are likewise deservedly counted amongst not Natural things. For whatsoever a Man may apprehend by speculation and reasoning, that is Supernatural; on the contrary, whatsoever he may handle, see, and feel, that is Natural.

Thirdly, the third part of things Natural and Supernatural, does consist in the Medicine or Physick, arising out of every one's hidden power and virtue; which Medicine made of anything, must be expelled out of a Visible, Palpable, and Natural body; and be reduced into a Spiritual, Exalted, and Supernatural Operation: that so the Spirit, which in the beginning was infused into, and granted to that body for its life, may be unlocked, and rendered penetrant to work, as a spiritual Essence and Fire, to which there are left its Vent-holes to burn, and so finds no obstacle that might choke, suppress, or hinder the burning life: Otherwise where the separation of the Soul and Spirit from the Body does not go before, there can no operation of any efficacy or advantage follow, according as necessity requires: For whatsoever is visible, palpable, and unseparated bodily, is natural and corporeal, but as soon as the separation is made, that which is quick leaves the dead, recovers its perfect operation, and, because the natural body is separated the spiritual Essence is loosened and freed to penetrate, and is become a spiritual Supernatural Medicine.

To sum up all, all things, nothing excepted, that may be handled and felt are Natural, but they must be made Supernatural, in case a Medicine be prepared out of them. For the Supernatural alone has in it a lively and quick virtue to work, but the

Natural has but a dead palpable form. For when *Adam* was Created, he was dead, and had no life of any virtue; but as soon as the operating, quickening spirit entered him, he showed his lively virtue and power by Supernatural admiration. In everything therefore, both Natural and Supernatural are copulated as one, and joined together in their habitation, that everything may be perfect. For of all things created in the whole World, some are Supernatural, as concerning only things Spiritual and of the Soul: but some are Natural and Supernatural together, as concerning the Elements and the firmament; and such are the Minerals, Vegetables, and Animals: Which is discerned and found so to be, when these are separated one from the other, that the Soul goes out of the body, and the Spirit forsakes its soul, and leaving the body an empty habitation.

You are furthermore to understand and to observe, that both the great and the little World are made, formed, and created out of one and the same first Matter, by an infinite and omnipotent essence, in the beginning at that time, when the Spirit of God moved upon the Water, who had been from eternity without beginning. The great World, that is, Heaven and Earth, was made first; and then the little World, which is Man, was made and formed out of the great, and the Water separated from the Earth. The Water was the Matter, upon which the everlasting Spirit of God moved: out of the noblest Earth, as its quintessence, was formed the little World, through the aquosity, which as yet was with the Earth, and all this was only Natural; But after the inspiration of the warm Divine breath, there came presently the Supernatural to it, that so Natural and Supernatural were united and copulated. The great World is transitory, though there shall be a new Earth or World, but the little World is everlasting. The great transitory created World will be reduced into nothing, as it has been made out of nothing: But the little World will be clarified by the Spirit of Ged, because he has the possession of it, and will make out of the first terrestrial Water, a heavenly clarified Water. Then will follow, that the first Matter has been changed into the last, and the last Matter is made the first again. The reason why the great World is transitory, is, because the Spirit of God does not keep his seat and habitation in the great, but in the little World: For the Man is the Temple of the Holy Ghost, except if he maliciously defiles himself to Hell-fire; that so there might be a difference and that Spirit remain in the little World, which he has formed after his own Image, and made it a Holy Temple. Besides this, there are all things in the little World, which are found in the great, *viz*. Heaven and Earth, together with the Elements, and what belongs to them of the Firmament.

We do find likewise, that in the first Creation, which was out of nothing, there did arise three Things; an Essence like to the Soul, and a Spiritual, and a Visible one, which did present a Mercurial Water, a sulphureous stream of Brimstone, and a terrestrial Salt. These three gave a complete and perfect, palpable and formal body to all things, wherein especially are found all the four Elements perfect. But of this I have already made mention in my Writings, but more particularly in a book of mine *de Microcosme*.

But to speak something more of Natural and Supernatural and Corporeal we do find, that the Woman of *Canaan* was cured of her Bloody Flux, which had lasted twelve years, by a touching only, having only touched the Garment of the Son of God; which distemper of hers was Natural, but the Medicine was Supernatural, because she had obtained this remedy of Christ the Lord by her faith. In like manner do we see an excellent, high Supernatural miracle in those three Men, *Sadrach*, *Mesach*, and *Abednego*, who being cast into the fiery Furnace, by the commandment of King *Nebuchadnezar*, were miraculously without any hurt delivered by Gods mighty power, *Dan.* 3.

Thus likewise the confusion of Tongues, and infusion of several languages, which happened at the foolish Building of the Tower of *Babel*, which should reach to Heaven, has been reputed a Supernatural Wonder and Miracle. Furthermore, it was a Supernatural Sign, that those *Israelites*, which should fight against the Army of the *Midianites*, according to Gods command, must lap of the Water with their Tongues as a Dog laps. And the sending abroad of the Dove out of the Ark of *Noah*, which brought with her a green Olive-leaf, as a token of mercy, was a Divine and Supernatural Message. That the Holy Man of God *Moses* did strike with his Staff the hard Rock, and the Waters gushed out of it at this stroke does transcend human reason. And it is no less Supernatural, that the Saltwater must become a sweet Water fit to be drunk. And likewise, the dry passage of the *Israelites* through the Red-Sea, and that *Aarons* staff did blossom are all Supernatural things.

To sum up all, that Christ the Eternal Son of God did raise himself out of the Sepulcher through the sealed stone: That he appeared to the two Disciples that went to *Emaus*: And that he manifested himself to his Disciples, the Doors being fast and shut, these are things Divine and Supernatural. I could allege more Examples out of the Holy Scripture, but I shall pass them over for brevity's sake.

Amongst Supernatural things are likewise numbered all such signs as happen in Mines, of bodily Apparitions of Spirits, Images, Fairies, and Dwarfs, which in several kinds visibly do appear, and do prognosticate either good or bad success, ruin or riches: As also all the Figures of Metals, and other shapes which are found in the Earth, of Men, Fishes, and other Beasts, framed and formed through the imagination of the three principles, but digested and ripened by the Earth and other Elements. Whereunto do belong likewise the Monsters of the Earth, and such like things as are found in the Earth at some certain times, of a wonderful form and shape, after the expiration of which times they are not to be found, and yet on another time they do appear again. To these we may refer all such shapes as are seen by means of Water,

Looking-glasses, which yet are of a different nature. For some are Natural only, and yet yield Supernatural shapes; But other some are produced by Conjuration, which are neither Natural nor Supernatural, but devilish; and therefore, they belong to Witchcraft, and pious Christians are prohibited to practice them: As likewise all such means, as are contrary to the Holy Scripture, Gods Word and Commandments, are justly rejected and refuted by the true natural Cabalists. I do speak this, that you nay make a due difference, and a certain order of Natural, Supernatural, and not Natural things.

Of the number of Supernatural things are likewise all water Spirits, as *Syrens*, *Succubi*, and other water Nymphs, and what belongs to them. As also the Spirits that are inhabitants of the Earth, and those of the Fire, which are heard, seen, and perceived; which bring sometimes tidings of death, or some other mischance; and sometimes do show, in certain places, riches and treasure, by Apparition. To which you may add the Spirits of Fire, which do appear in the shape of Fire, or form of a burning Candle; which indeed are all Spirits, and have impalpable bodies: but they are not such Spirits as those Spirits of Hell, which do hunt after Men's Souls, as the most precious jewel, as hellish Lucifer, the Devil and his accomplices do, which have been cast out with him: But they are such Spirits, which above Nature are the objects of Men's admiration, and live only by the Elements, and are sustained and fed by them, and with the dissolution of this Terrestrial World, will be annihilated and vanish away, because there is no redeemed Soul in them . I shall not insist longer upon this, but leave a further declaration of these circumstances for another occasion, when I shall give you a singular account of Visions, and Shapes, and Spiritual Apparitions, which by the greater part of the World are accounted to be not Natural, and yet are truly Natural, but they are found Supernatural in their operation and wonderful quality.

For the further confirmation of my purpose, I do say, that there are many things to be found in Medicine, which do show forth and perform their operation Supernaturally, after a magnetic manner, working only by an attractive spiritual power, which is attracted through the Air; because the Air is the *medium* betwixt the Medicine, and the disease or distemper. Like as the loadstone always longs for, and turns itself to its Star, though this Star be many thousands of miles distant from it, yet is the spiritual operation and affection so powerful betwixt these two, that they are drawn together by that *medium*, the Air, at so great and vast a distance. But because this attractive Power is generally acknowledged by all Men, it is grown a mere Custom, and so it is held, and nothing more of abstruseness observed, what the original is of this operating power.

In like manner, distempers and sores may be cured and healed, though the Patient and the Physician be far enough distant one from the other, not by

Benedictions or Conjurations, or other unlawful prohibited means, which are contrary to God and Nature; but by such means, wherein there is a magnetic attractive power to perform such things. As when a Patient goes away, and leaves the Weapon wherewith he has been wounded, or some of the blood which issued forth from the wound, with the Physician; if he does proceed orderly using the right means, as one uses in binding up and dressing of a wound, he will certainly recover his former health. This is no Witchcraft, but this healing is performed by the attractive power of the Medicine, which by means of the Air is conveyed to the wound, and thereby cleansed, for the performing of this spiritual operation.

These expressions will seem hard to many, and impossible to Nature, and many will say this relation is against Nature, whereby they will be moved to dispute and to argue this question, whether it be Natural, or no? Whether it be possible or no? Or whether this cure be not Witchcraft?

I shall compose the difference thus: That this cure is Natural, but the operation of the cure is Supernatural, because it is performed only in an attractive incomprehensible manner. And that this manner of curing or healing is no Witchcraft, I do demonstrate thus, because it is not mixed with any Witchcraft, nor any other means, which are either unnatural, or contrary to God the Creator, and his holy alone saving word; but being only Natural, out of their Supernatural, invisible, incomprehensible, spiritual, and attractive power, which has its original from the Stars, and performs its operation through the Elements.

Lastly, that this cure is no Witchcraft, I do prove thus, because the Devil takes rather a delight and pleasure in the sad mischances of mankind, then that he should administer any help for the welfare of Men: which besides he cannot possibly perform without Gods providence and gracious permission.

Much more might be written of this magnetic form, but to prevent divers errors, I will let it alone, till I come to speak something *de Miraculis Naturalibus*, or Natural Miracles of the World.

Those gross and silly Head-pieces, which nevertheless account themselves to be very wise Masters of Philosophy, and all such as have not the perfect use of their senses, cannot find any difference in these things. But he that is wise and understanding, knows to distinguish the Natural from the Supernatural. For do but observe and consider this similitude, to demonstrate the truth of this thing by a rude example, that there are found many beasts, which die in the Winter, and lie dead, so that there is no life to be discerned in them: But as soon as the warm Summer does draw nigh, the natural heat gives them a new life, that the reputed Carcass is fully revived in the same substance, it has had in its living motion. Like an Herb, which dies in Winter, and does appear new and fresh again in the Spring. Now the dying of such things is to be counted Natural, but the restoring them to a new life is in its knowledge Supernatural. But because men are used to these things, therefore the least part of them does regard that which deserves a further enquiry, and give over the thoughts of things both Natural and Supernatural.

Moreover, the greatest part of Men do pass by inbred and natural dispositions, which are likewise Supernatural; as also abortions, and such as bring some tokens with them into the World, which indeed are Natural, but through the occasioned imagination show themselves Supernatural, which supernatural form, and supernatural impressions have been produced by the Mother of the Child through thoughts arising, which unawares, and as it were by accident have happened unto her. As we do see and find many a time, that many Men have naturally inbred gestures, which they are never able to leave, though they endeavor it never so much. This inbred thing is Natural, but the conception in the Mother's womb, caused by the imagination of such a thing, is Supernatural, and subject to that which is made by the impression of Heaven.

Finally, if you will say that it is not true, that anyone can defend that which is Supernatural, with certain grounds and reasons, except he has learned the Natural, which has its original and form from the Supernatural; yet he will (after he has studied it) by a certain invented experiment be capable to demonstrate that he is a conqueror of these that will not believe that which is Supernatural, and confute those that do presume to dispute of Natural things, and because they are ignorant of the foundation, do nothing but talk idly, and quarrel unprofitably.

Chapter II. Of the first Tincture, the Root of Metal.

To return to my purpose and undertaking, which is to discourse by Gods permission, and to give an account of the first Tincture, the Root and Birth of Metals and Minerals; it is to be observed, that the Tincture, which is the Root of all Metals, is likewise a Supernatural flying fiery Spirit, having its sustenance in the and[24] looking naturally for its habitation in the Earth and Water, where it may rest and work. And this Spirit is found in all Metals, and more abundant in other Metals than Gold: For the Gold is very close, solid, and compact, by reason of its well digested, ripened, and fixed body; therefore it can no more enter into the body, than the body does need. But other Metals have not such a fixed body, but their pores are open and dispersed, therefore can the tinging Spirit abundantly more penetrate and possess them; but because the bodies of other Metals are unfixed, the Tincture likewise cannot stay with these unfixed bodies, but must go out of them. And being the

---

[24] It may be that there is a word missing in "the and", but it does not appear in the printed edition of 1671. -pnw

Tincture of Gold does in no other Metal abound more than in Iron and Copper, as Husband and Wife, their bodies are destroyed, and the tinging Spirit from thence expelled, which breeds much blood in the opened prepared Gold, and by its feeding does make it volatile. Therefore when the volatile Gold is filled by its meat and drink, it takes up its own blood, does dry it up through its own internal fire, with help and addition of a moist fire, and is again a conquest, which does fix, nay, produces the highest fixedness, so that the Gold becomes a high fixed Medicine, and cannot make a body again, by the reason of the superabundant blood, except there be added to it a superfluous body, into which the abundant fixed blood does disperse itself, which joined metallic body is penetrated by the exceeding great heat of the fixed blood of the Lion, like fire cleansed from all impurity, and immediately ripened to a perfect maturity and fixedness: That thus the servant enriches first his Lord and Master, because the Lord cannot spare of his cloth to give away, by reason that Nature has granted him but one only Suit of Honor; and the King can distribute again the Inheritance, and Court-cloths of his Kingdom to his Servants, after he has first taken and gathered the Tribute of his subjects; that so the Master and Servant may stay and continue together. Nor need you to wonder at this, that the King must borrow of his servants, because their bodies are not fixed and permanent, for they take up much, and can keep but small credit. But if the King can participate of it, he can better overcome heat and frost, then the leprous Metals. And thus, he becomes by this participation a Regent and Conqueror, particularly of all others, with a great victory and triumph of riches and health to a long life.

I hope you have understood a:rd learned enough for the beginning of this discourse of Natural and Supernatural things, and the first Tincture the Root of Metals and Minerals, whereupon the Corner-stone is placed, and the true Rock in general is founded, wherein Nature has placed, and concealed, or buried her abstruse and deep hidden gift, *viz.* in the fiery and tinged Spirits, which Tincture they got from the Starry Heaven, by the operation of the Elements, and are made able further to tinge and to fix, that which had no tincture, and was unfixed before: because that *Lune* wants the suit of the golden Crown, together with the fixedness, as also *Saturn*, *Jupiter*, and *Mercury*. And though *Mars* and *Venus* do not want this clothing, but may communicate the same to the other five; yet do I say, that they can do nothing for the gaining of riches without the Lion, because they are not for their need, provided with the fixedness of their Mercury, and the flexibleness of their Salt; except the Lion has overcome them in the fight, and they are notably mended and bettered which melioration lies hidden in the signate Star, or their Load-stone, from which all the Metals have received their gifts.

I do proceed, and in Specie pass to the birth and to the generation, how the *Archæus* does show and pour forth its power, and displays it, by which all the Metallic and Mineral forms are exposed to the view, and are made formal, palpable, and corporeal, through the Mineral, incomprehensible flying, fiery Spirits. Furthermore, you are to know, and with all diligence to observe, not to pass by with oblivion that which is of great concernment, nor to look over that which is most advantageous, and on the contrary to note all along the mere writing, not regarding the drift and scope: For of that I do write here, the highest will be undoubtedly accounted and esteemed by many the lowest, and the slightest the highest mystery.

First of all, you are to know, that all the Metals and Minerals of the Earth have one only Matter, and one only Mother, by which they in general altogether have received their conception and perfect bodily birth. And this Matter, which comes from the Center, does divide itself in the beginning into three parts, to produce some corporeal thing, and a certain form of every Metal. These three parts are fed and nourished by the Elements in the Earth out of its body, till they become perfect. But the Matter, which has its original from the Centre, is framed by the Stars, wrought by the Elements, formed by that which is Terrestrial, and is a known Matter, and the true Mother of Metals and Minerals: and is such a Matter and Mother, out of which Man himself has been conceived, born, nourished and made corporeal: And may be altogether compared to the middle World; for whatsoever is in the great World, that is likewise in the little, and whatsoever is in the little World, that is likewise in the great: And thus what is in the great and little World together, that is found likewise in the middle World, which joins the great and little World, and is a soul, which does unite and copulate the spirit with the body. This Soul is compared to Water, and is indeed a right true Water, yet does not wet like other Water, but it is a heavenly Water, found dry in a Metallic liquid substance, and a Soul like Water, which loves all Spirits, and does unite them with their bodies, and brings them to a perfect life. Therefore, it is certain, that the Water is a Mother of all Metals, which being heated by a warm Aereal fire, as is the Spirit of Sulphur, brings life into the Terrestrial body through its ripening, wherein the Salt is apparently found, which does preserve from putrefaction, that nothing may be consumed by corruption. In the beginning, and in the birth is wrought first of all the Quick-silver, which yet lies open with a subtle coagulation, because there is but little of the Salt communicated to it; whereby it shows more a spiritual, than a corporeal body. Other Metals, which are derived from its Essence, and have more Salt, which makes them corporeal, do follow after this. I begin with the Spirit of Mercury.

Chapter III. Of the Spirit of Mercury.

Although I do use a peculiar style in my Writings, which will seem strange to many, yet there is sufficient cause for it. It is sufficient, I say, to insist upon my

Experience, and not to regard other men's idle speeches, because I have attained to the knowledge of these things, and seeing goes always before hearing; and that which has its foundation is preferred before that, which has no ground at all. Therefore, I say, that all visible, and palpable things are made of the Spirit of Mercury, which is beyond all the Terrestrial things of the whole World, and all things are made of it, and have their original from it. For therein is all to be found, that can do all, what the Artist does desire to enquire into. It is the principle to work Metals, being made a Spiritual Essence, which is a mere Air, and flies to and fro without wings, and is a moving wind, which after its expulsion out of its habitation by *Vulcan*, is driven into its *Chaos*, into which it enters again, and does resolve itself into the Elements, where it is attracted by the Stars, after a magnetic manner, out of love, from whence it went forth, and was wrought out before, because it desired to be united again with its like. But when the Spirit of Mercury can be taken, and made corporeal, it does then resolve itself into a body, and becomes a clear, fair, and transparent Water, which is true Spiritual Water, and the first Mercurial Root of Minerals and Metals, Spiritual, unperceivable, incombustible, without any commixition of the Terrestrial aquosity. It is that heavenly Water, of which much has been written, for by this Spirit of Mercury all Metals may be, if need requires, dissolved, opened, and without any corrosive reduced or resolved into their first Matter. This Spirit renews both Man and Beast, like the Eagle; consumes whatsoever is bad, and produces a great Age to a long life. This Spirit of Mercury is the chief Key of all my other Keys, of which I have written in the beginning. Therefore will I call; Come ye blessed of the Lord, be ye anointed with Oil, and refreshed with Water: Embalm your bodies, that they may not putrefy, get a bad scent and stink: For the heavenly Water is the beginning, and the Oil a medium, which does not burn, because it is made out of a spiritual Sulphur; and the balsome of Salt is corporeal, which is united with the Water by means of the Oil: whereof I shall give you hereafter a more ample account, where I do intend to speak and to write, something more concerning these things.

And to declare further the Essence, matter and form of this Spirit of Mercury; I must tell you, that its Essence is Soul like, its Matter Spiritual, and its form terrestrial, which yet must be understood by some incomprehensible thing. These are indeed hard words and expressions, and there are many that will think, that these are vain and idle repetitions, and strange sayings, which do produce nothing else but strange thoughts. It is true, I confess, they are strange, and do require strange people that will understand the meaning of them. It is not a thing written for Country-men, how they must grease their wagons: Nor is it a speech for those, which have not got the knowledge of this Art, though they imagine themselves to be never so wise. But this man alone I do repute to be learned, which besides the word of God, does make enquiry by a true knowledge into Terrestrial things, which come under the judgment of reason, and learns to know the darkness out of light, and to choose the seeming bad before the good.

Touching the beginning of this Spirit of Mercury, this is needless to know, because it is of no benefit, nor can it do you any good. But observe, that its beginning is supernaturally from Heaven, the Stars, and Elements, granted in the beginning of the first Creation, to enter further into a terrestrial being. And because this is needless, as I have told you, leave that which is Heavenly to the Soul, and apprehend it by Faith; that which is of the Stars, let likewise alone, because such impressions of the Stars are invisible and incomprehensible; the Elements have already brought forth this Spirit perfect into the world, through the nourishment of it, therefore do not meddle with them either; for no man can make any Element but the Creator alone, and insist upon thy Spirit already produced, which is both formal and not formal, comprehensible and incomprehensible, and yet does appear visibly, and you have the first Matter, out of which are grown all Metals and Minerals, and is one only thing, and such a Matter, which does unite itself with the Sulphur of the next following Chapter, and is coagulated with the Salt of the Fifth Chapter, so that it becomes one body and a perfect Medicine of all Metals, not only to generate in the beginning in the Earth, as in the great World, but also, with help of a moist fire, to change and transmute together with the augmentation in the little World. Let this not seem strange to you, because the most high has thus permitted it, and Nature has wrought it.

There are many in the World which will not believe this, and do think it impossible, that vilify and despise these mysteries, which they in no wise understand. Those may continue in their folly like Asses and Blockheads, till they are illuminated, which does not happen without the will of God, but comes by his Providence. But wise and experienced Men, which have wrought in the sweat of their Brows, will bear me witness, and confirm the truth; and likewise avouch, that they really believe and think, that whatsoever I do write here, is nothing but truth, as true as Heaven and Hell is made and ordained for the elect and damned, for a reward of good and evil. I do not write with my hands only, but my heart and Soul does compel and urge me to do it, because that many conceited, illuminated, reputed skillful worldlings do hate, envy; disparage, rail at, and persecute this mystery to the outmost skin, or to the inmost kernel, which has its original from the center. But I am sure, that time will come, when my marrow is varnished, and these bones of mine are dried up, that some people will sincerely take my part, though I am in my grave, and would be willing to fetch me from the dead, if God would permit it, but that will be a thing impossible. Therefore, I have left them in writing, whereby their faith and confidence will have a seal of certainty and truth, to bear witness of me, what has been my last Will and Testament, which I have left to the poor, and to all the admirers of mysteries. Though it did not become me to write so many things, yet could I not, without doing hurt to my Soul do otherwise, then to drive a glance and brightness through the clouds, that the day may appear, and the obscure night, together with the cloudy and dark tempests may be dispersed.

But how the *Archaeus* works further by the Spirit of Mercury in the Earth, or in the veins of the Earth, you are to understand, that after the Spiritual Seed is framed from above by the impression of the Stars, and fed and nourished through the Elements, this Seed is changed into, and is become a Mercurial Water: as in the beginning the great world likewise was made of nothing; for the Spirit moved upon the Water, and thus was this cold, waterish, and terrestrial Creature revived to life by a heavenly warmth. It was in the great World the Power of God, and the operation of the light of Heaven; in the little World, likewise the Power of God, and the operation by his divine and holy breath to work in the Earth. Furthermore, the Almighty did grant and ordain means, for the performing of the same, that the Creature might get power to work upon another creature, and one might help and promote the other, for the performing and perfecting of all the works of the Lord. Thus was granted to the Earth an influence to generate by the Luminaries of Heaven, and likewise an internal heat to warm & to ripen that which was too cold for the Earth, by reason of its aquosity, and thus every Creature a peculiar Genius according to its kind; that so there is raised a subtle sulphureous steam by the starry Heaven; not a common, but another clarified, clean, and pure steam, separated from others, which unites itself with the Mercurial substance; by which warm property in a long time the humidity is dried up by little and little, and then the Soul-like property being joined with it, which gives the body and balsome of maintenance, and works before too upon the earth by a spiritual and starry influence. Thus, happens then a generation of Metals, according to the commixtion of the three principles, and according as they take in more or less of these three, so the body is formed. If so be the Spirit of Mercury is directed and formed from above upon Animals, then is there produced an Animal Being; but if it seizes upon Vegetables, a Vegetable work is brought forth. And if it falls upon Minerals, by reason of its infused nature, there will spring thence Minerals and Metals. Nevertheless, everyone is differently wrought: The Animals by another form by themselves, the Vegetables after a manner proper to themselves, and the Minerals likewise in another fashion, every one after a singular way: Whereof in specie much more might be written, and a more full and exact relation might be made.

Here may be justly demanded, how this Spirit of Mercury may be had and obtained, or how it is to be made, and which way, and after what manner it may be prepared, that it may cure distempers, and change and alter all Metals of the ignobler kind, as they are generated in the little World, by a transmutation and augmentation of their seed? Many will expect an answer to this question, which I shall not keep back from them, but faithfully discover, as much as I have leave by Gods command and judgement, in the manner as follows.

Take in the Name of the Lord red Mineral quicksilver, which looks like Cinnaber, and the best mineral Gold that can be gotten; take an equal quantity of them both, and grind them together, before they have been in any fire, pour upon then an Oil of Mercury made by itself, out of the common putrefied and sublimed Quicksilver; digest them for a month, and you will have an Extraction, which is more heavenly than terrestrial. Distill gently this Extraction in *Balneo Mariæ*, and the phlegm comes over, and the ponderous Oil remains in the bottom, which takes up unto itself all Metals in a moment. Add to this three times the quantity of Spirit of Wine, circulate it in a Pelican till it becomes blood red, and has recovered an incomparable sweetness. Pour off the Spirit of Wine, and add to it fresh Spirit of Wine. Repeat this so long, till the whole matter be dissolved into an exceedingly sweet and Ruby colored transparent liquor, which mingle afterwards together. Pour it upon white calcined Tartar, and distill it with a strong fire in Ashes, and the Spirit of Wine remains behind with the Tartar, but the Spirit of Mercury comes over. This Spirit of Mercury being mixed with the Spirit of Sulphur Solis, together with its Salt, whosoever shall bring them over thus joined and united together, that they may not be separated in *infinitum*, he will have such a work (if so be it does receive its ferment in a due measure and prefixed term, with Gold by a Solution, and is brought in its perfect maturity to a plusquam perfection) to which nothing nay be compared, for the preventing of diseases and poverty, and for a rich and superfluous recreation of the body as well as of goods.

This is the way to obtain the Spirit of Mercury, which I have discovered so far, as the Highest Emperor has given ne leave to do. I hope you will use wisely and discreetly my Manual Operations required to do this work, and laid open by me, that you may not suffer in Hell fire for my faithful warnings; because the door which gives entrance to the Kings Court, is unlocked and opened by one Key alone, which cures all distempers, as the Dropsie, Consumption, the Gout, the Stone, the Falling-sickness, Apoplexy, Leprosie, and of what name soever they be in general. This is likewise a remedy for all sorts of French-pox, and all other old lasting sores, as the Wolf, Tetters, Worm, Fistulas, Cancer, Spreading and fretting Ulcers and Holes, as I have discovered too, and hidden nothing from you.

Finally, observe this, that you will make known only this, and no more, that, because any Art has its beginning and original from the Spirit of Mercury, which is quickened and revived by the spiritual Sulphur, that an heavenly thing does rise from them together, and with, and by the Salt they become corporeal and formal; but the principal of the Soul , the Spirit, and the Body, you will let it be and continue a Loadstone, as really it is, and cannot be accounted to be any other thing. But the final sum is this, that without the Spirit of Mercury, which is the only Key to make the corporeal Gold Potable, the Philosophers Stone can never be made nor prepared. Do not object against, nor contradict this sentence, but keep silence: For I will give over speaking, because silence is imposed upon you and me by the competent judge, and begin the Execution yourself, leaving further enquiry to another, which has not as yet pleaded his cause.

Chapter IV. Of the Spirit of Copper.

The Star named *Venus* is hard and very difficult to be reckoned, as all the Mathematicians and Astronomers must bear me witness; for her course does much differ from the other Six Planets, therefore her Nativity is likewise of another nature; because the birth of *Venus* does possess the first table after *Mercury*. As concerning the generation of Metals, *Mercury* makes effectual, but *Venus* does incite, and gives lust and desire, together with beautifulness, which does occasion it. Though I do not esteem myself, nor take upon me to be an Astronomer, to whom the account of the heavenly motion is known, because I ought to spend my time in prayers in the House of God: Nevertheless, that the remainder of the time, after Divine service is performed, may not pass idly away, I have resolved to spend the rest of my time in enquiring into natural things. Thus, it is a hard matter to find out what is produced and brought forth by *Venus*, or whence *Venus* has taken her beginning and original, because she is clothed to excess, with that she has no need of; and on the other side must want, that she stands in need of, as touching her fixedness.

But you are to know, that *Venus* is clothed with a heavenly Sulphur, which does far exceed the Splendor of the Sun, because there is found much more Sulphur in her, than in Gold. But that you may learn, what the matter is of the said Sulphur of Gold, which dwells and reigns abundantly in *Venus*, and of which I have spoken so much, know then, that it is likewise a flying and very hot Spirit, which can search and penetrate all, and also digest, ripen, and bring to maturity, *viz.* the imperfect Metal into perfect, which the unexperienced does not believe. If you ask, how the Spirit of Copper can ripen, and bring to perfection other imperfect Metals, it being itself, in its body imperfect, and not fixed? I answer, as I have told already, that this Spirit cannot have or hold in Copper a fixed body for a habitation; therefore the habitation being burned by fire, the guest goes out of it likewise, and must leave his habitation with impatience, for he dwells therein like a hireling. But in the fixed body of Gold he has a protection, that nothing can drive him out without the Sentence of a peculiar judge, because he has taken possession of his habitation like an heir, and has taken root in that fixed body, that cannot be cast out so easily. The Tincture, which *Venus* has obtained, is likewise to be found in *Mars*, yea, much more powerful, higher, and more excellent: For *Mars* is the Husband, *Venus* the Wife; whereof I have spoken more at large in those Writings of mine concerning them. This Tincture is likewise found in Verdigrease and Vitriol, as in a Mineral, of which a whole Volume might be written: And in all these things there is found a Sulphur, which does burn, and yet another Sulphur, which does not burn, which is a wonderful work. The one is white, the other red in the operating birth: But the right and true Sulphur is incombustible: For it is a mere and true Spirit, out of which is prepared an incombustible Oil, and is indeed the Sulphur out of which the Sulphur of Gold, out of one and the same root, is made and prepared. I do discover many secrets, which ought not to be done; but I do not know how to help it. To conceal all, is likewise a thing unanswerable, yet it is good not to do overmuch; as I have desired in that Protestation of mine, not to forget my request.

This Sulphur may very well be called and christened the Sulphur of the wise, because in it is found all wisdom, if you accept the Mercurial Spirit, which is to be preferred, and with it, together with the Salt of *Mars* must be united through a Spiritual copulation, that three may be brought to a correspondency, and be exalted into one operation.

This Spiritual Sulphur does likewise and in the same manner derive its original from the Upper Region, as the Spirit of Mercury does, but with another form and fashion, whereby the Stars do show a separation in fixed and unfixed, in tinged and not tinged things. The Tincture does consist only in the Spirit of Copper, and chiefly of its consort, and is a mere steam, stinking and of a very ill-scent in the beginning: And this mist must be resolved in a liquid manner, that the stinking incombustible Oil may be prepared out of it, which yet must have its original from *Mars*. This Oil is easily joined with the Spirit of Mercury, and does soon take up all metallic bodies, being first prepared according to the account given by me in my Keys.

I do not keep here any order of the Planets, and that justly, not without some reason; for I do follow their generation, that is, the order and rank which I do follow because *Venus* has much Sulphur, she has been together with *Mars*, digested and ripened sooner than any other Metals; but because they have had but little help from the inconstant Mercury, being he had no room left him to work harder, by reason of the superabundant Sulphur, they could not receive or obtain a melioration of their unfixed bodies. Now will I discover a mystery to you, that *Gold, Venus* and *Mars*, have in them one and the same Sulphur, one Tincture, and the same Matter of their Tincture, which Matter of the Tincture is a Spirit, a Mist, and Fume, as has been said before, which has penetrated., and does penetrate all bodies: If you can bring it into captivity, and do actuate it with the Spirit, which is found in the Salt of *Mars*, and then do join with the same the Spirit of Mercury, according to their weight, and do separate them from all impurity, that they become sweet, and sweet-smelling, without any corrosive, you have then a Medicine, to which nothing in the World may be compared; if you ferment this Medicine with the shining Sun, you have made an ingress, which is penetrant to work and to transmute all Metals.

O Eternal Wisdom, what thanks must be rendered unto thee, for these great mysteries, which yet the children of men do not regard at all, and do scorn to enquire, and to learn what you have hidden in Nature. They see it with their eyes, and do not know it; they hold it in their hands, and do not comprehend it; they touch and handle it, and do not know what they have, or what they make, because the inward part is concealed from them.

I will lastly really, out of love to God, discover yet this unto you, that the root of the Philosophical Sulphur, which is a heavenly Spirit, together with the root of the spiritual supernatural Mercury, and the principle of the supernatural Salt, is in one, and is found in one Matter, out of which the Stone, which has been before me, is made, and not in many things; although the Mercury be drawn by itself by all the Philosophers, and the Sulphur by itself, besides the Salt apart. That so Mercury is found in one, and the Sulphur in one, and the Salt in one. Notwithstanding all this, I do tell you, that this is to be understood of their superfluity, which is found most in every one, and particularly in many ways may be used profitably, and prepared to a Medicine and transmutation of Metals. But the Universal, as the greatest treasure of terrestrial knowledge and wisdom, and of all the three principles, is one only thing, and is found in one only thing, and drawn out of it, which can reduce all Metals into one only thing, and that is the true Spirit of Mercury, and the Soul of Sulphur joined together with the Spiritual Salt, enclosed under one heaven, and dwelling in one body, and is the Dragon and the Eagle, it is the King and the Lion, it is the Spirit and the body, which must tinge the body of Gold to be a Medicine, whereby it gets abundant power to tinge others its consorts.

O blessed Medicine, granted by God thy Creator! O heavenly Loadstone of that great attractive love! O bountiful substance of Metals, how great is thy power, how unsearchable is thy virtue, how stout is thy constancy? He is blessed upon Earth, that has got a real knowledge of thy light, which the world takes no notice of. He shall not suffer poverty, no distemper shall touch him, no disease shall do him any hurt, till to the prefixed time of death, and to his last hour, which the King of Heaven has set and prefixed. It is impossible for all the tongues of Men to express and to declare the wisdom, which is laid in the treasure of this fountain. All the Orators must become dumb, and be brought to confusion, nay be astonished and made incapable of speaking anything, if they should behold and know the Supernatural Majesty. And I am amazed myself, when I do think and consider, that I have revealed so many things; but I hope with my prayers to prevail with God, that he may not lay this to my charge as a mortal sin, because I have begun this work in his fear, have obtained it by his mercy, and have revealed it to his glory and praise.

O most Holy and everlasting Trinity, I do give unto thee both with my heart and mouth, praise, honor, and glory, for that you have revealed to me the great wisdom of the terrestrial World, besides thy Divine Word, whereby I have got the knowledge of thy Almighty power, and Supernatural miracles, which men will not acknowledge. I do most humbly beseech thee to grant unto me further prudence and knowledge, to make use of their power and virtue with perpetual thanksgiving to thee, to the benefit of my neighbor, and to my own welfare as well Spiritual as Corporal. That so thy Name may be praised, magnified, and glorified, for all thy creatures both in Heaven arid on Earth, and my enemies may acknowledge that thou art a Lord full of infinite wonders, that they likewise once may repent and be converted, and not perish in the darkness of falsehood. So help me and us all God the Father, God the Son, and God the Holy Ghost, exalted above all in his Throne, Glory, Power, and Majesty; whose wisdom has neither beginning nor end, and before whom all the Creatures, heavenly, terrestrial and hellish, with fear do shake and tremble, blessed and praised for evermore, *Amen.* O Seraphin, O Cherubin, how great are thy wonders and works, look in mercy upon thy Servant, and turn thine anger from him, because he has revealed these things.

Concerning the generation of Copper, the reader is moreover to know and to observe, that the Copper is generated out of much Brimstone, but its Mercury and Salt are equal in the same, for there is neither more nor less in quantity of one and the other to be found. Now because the Brimstone does exceed in quantity the Salt and Mercury, there arises from thence a great tinging redness, which great redness, has so possessed the Metal, that the Mercury could not perfect its fixedness, that a more fixed body might have been produced out of it. You are further to know, that the form of *Venus* body is of the same condition that a Tree is of, which has and does yield abundance of gum, as is the Pine and Fir-tree, with other sorts of trees more, which gum is the Sulphur of the tree, which drives out sometimes this gum at the sides of it, by reason of its too great abundance, and because it cannot harbor it all. Such a tree now, that is tinged with so much fatness by Nature, and the ripening of the Elements, burns and takes fire immediately, neither is it heavy, and is never so durable as Oak, and the like hardwood, which is solid and compact, and has not its pores so open, as that sort of light wood, that the Brimstone might abundantly reign in it. But therefore, has the Oak-wood more Mercury, and a better Salt, than the Pine or Fir-tree. And such wood is never so much apt to swim upon the water, as the Fir-tree is, because it is close, solid, and compact, that the Air in it cannot bear it up. The same is to be understood of Metals, but especially of Gold, which, by reason of its much fixed and well ripened quick-silver, has a most solid, compact, close, fixed, and invincible body, to which neither fire or water, neither air, nor any putrefaction of the Earth can do any hurt, because its pores are closed-up, arid the corrupting power of the Elements cannot injure it. Which fixedness, and solid, and compact conjunction do demonstrate its natural ponderosity, which is not to be found or proved in other Metals, which may be discerned not only by weighing it in a pair of scales, but you will find it likewise, if you put but a Scruple of pure gold upon a hundred pound weight of Quick-silver, it will fall presently to the bottom, whereas all other ponderous Metals, laid upon Quick-silver swim upon it, and do not sink to the bottom, because their pores are more largely extended, that the Air or Wind may pass through them to bear them up.

Furthermore, concerning the Spirit of *Venus* or Copper in Physick, you are in fine to observe, that it is found very necessary and wholesome in its virtue and efficacy, not only that Spirit which lies in *Primo Ente*, but that Spirit likewise which is found in the last Matter: Its virtue, power, and operation is such, that in rising of the Mother it is to be preferred before any Medicine whatsoever; as likewise against the Falling Sickness particularly there is nothing comparable to it. This Spirit has moreover received a singular gift to dry up the Dropsie. It preserves the blood from putrefaction, and does digest everything, that might be against, or be hurtful to the stomach. It breaks the Stone, of whose nature it is. Outwardly, it lays the foundation in wounds for the cure of them. The sore called long ago *Noli me tangere*, and any old ulcers, be they never so deep rooted, it lays hold upon their malignity, and ushers in a ground for the healing of them. Outwardly it does purify, and searches for the certain kernel, where the Cure and working Medicine may fasten, and have its beginning. But inwardly it searches and penetrates thoroughly, and finds out any malignity in the body of Man. It is like to the noblest wound-drink; there is no Imposthume, but it is cured by it.

To sum up all, I do say, if you have a special care of this Spirit of Copper, it will work such wonders both inwardly and outwardly, as will be accounted of all incredible and supernatural. And thus much of the Spirit of Copper.

Lastly, and to conclude all, you are to know, that the Spirit of Copper is a hot Spirit, penetrant and searching, consuming all the bad humours and phlegm, both in Men and Metals, and may be justly accounted the Crown of Physick. It is very fiery and piercing, incombustible, yet Spiritual and without form; and therefore, is capable like a Spirit to further in particular the ignition, digestion, and ripening of things without a form.

And if you are a true enquirer into Natures *Arcana*, let this Spirit be recommended unto you, it will never forsake you in any necessities, or wants either of health or riches, if you do exactly observe, and justly administer it. I am in hopes my requests and desires will once find place with, and be heard of many which make enquiry into Nature, and are very desirous to search after and to learn its secrets. Therefore, they will. whet their senses, open their eyes, and give leave to their ears to hear, that such things may be learned out of this relation of mine, that never were observed, nor rightly understood before, which are found in this Spirit of Copper both inwardly and outwardly. He that will not give heed to nor observe and understand these writings of mine, has not found out many mysteries, nor enquired with constancy and truth without me, neither learned or got any profitable knowledge. Therefore, no man can pass his verdict upon me, as touching the Spirit of *Venus*, except he has turned the Copper, and exactly studied all the secrets of its inward virtue, as I have done. If I can get the knowledge of anything that is better, which I am as yet ignorant of, I do most earnestly beseech you, not to conceal anything, and his doctrine shall be very well rewarded in a thousand ways. And thus, I recommend you to the most high Creator. Reason cannot always apprehend, that which *Venus* can reach into. No thoughts can quickly find it out, and human wit thrusts it far from itself. Its Spirit alone will judge all, and the Mercury will then co-operate with it.

Chapter V. Of the Generation of Mars, its Spirit and Tincture.

Mars and *Venus* have one and the same Spirit and Tincture, as the Gold and other Metals have, though this Spirit be found in every Metal, in some a greater, in others a smaller quantity, it is undeniable, and confessed of all, that there are divers men, and divers opinions; although men in the beginning are made out of one first Matter, and generated and born out of one Seed, yet is there a manifold difference of their opinions, because the operation of the Stars has occasioned this, and not without a cause; for the influence of the great World works the other, (namely the difference of opinions) after itself in the little World; because all the Opinions, Nature and Thoughts together, with the whole complexion of Man, do derive their original only from the influence of the Stars of Heaven, and do show themselves according to the Planets and Stars, where nothing can withstand, nor obstruct such an influence, because the generation of their perfection is already performed and brought to a period or finished. For example, a man is naturally inclined to Study, one has a mind for Divinity, another for the study of the Law, the third for Physick, the fourth will be a Philosopher. Besides all this, there are many wits, that have a natural inclination for Mechanic Arts; as one turns a Limner, another a Goldsmith; this man a Shoe-maker, that man a Tailor, another a Carver, and so forth, manifold and innumerable. All this happens by the influence of the Stars, whereby the imagination is strengthened and founded supernaturally, wherein it resolves to continue. As we do find, if a man has once taken up a resolution in his mind, and laid a foundation upon it, that no man is able to bring or keep him from it, that he should not so obstinately stand upon it, death only excepted, which at last closes up all. The same is to be understood of Chemists and Alchemists, which having got once into the secrets of Nature, do not intend to give them over so easily, except they have more exactly searched Nature, and wholly absolved and finished the study thereof, which yet is no easy matter. Thus, you are likewise to understand of Metals, that according as the infusion and imagination happening from above, so happens the form likewise, although Metals are altogether called Metals, and are indeed Metals, yet as you have understood by the divers opinions of men, which are altogether men out of one matter, there may be manifold and divers Metals, of which one has got a hot and dry, another a cold and moist, another a mixed complexion and nature. Therefore, because the Metal of *Mars* has before others been ordered by a gross Salt in the greatest quantity in its degree, its body is the hardest, most inflexible, strongest, and coarsest, which Nature has thus granted and appropriated to it. It contains the least part of Mercury, a little more it has of Sulphur, but the greatest part of Salt: And from this mixture is sprung its corporeal being, and is thus born into the World, with the help of the Elements. Its Spirit is in operation equal to other Spirits: But if the true and right Spirit of Iron can be discerned, I do really, and not unwisely tell you, that one grain of its Spirit or quintessence, taken and administered in Spirit of Wine, comforts and strengthens a man's heart, mind, and courage, so that no fear of any of his enemies may be perceived: It stirs up a Lion like heart within him, and does inflame to begin and consummate a fight with *Venus*. If the Conjunction of *Mars* and *Venus* does rightly happen in a certain constellation, they have success, victory, and conquest, both in love and sorrow, in fights and peace, and will continue of one mind, though the whole world should bear a spleen and enmity against them. But because I am an Ecclesiastical man, I have subjected myself to Spirituality, and have recommended my Soul to God, without enticing of human concupiscence, and allurements of the lust of the flesh, which being unpermitted, prepare a way to Hell; but Gods command, fear, and permission of the will of men, licensed by his command, make the way ready for Heaven, if they do persevere in the true worship of and the true lively faith in the Throne of Grace, the Mediator and Intercessor Jesus Christ our Savior, This Spirit cures, dispels and heals wonderfully all Martial distempers, as the Dysentery or Bloody-flux, the Courses of Women, white and red, any looseness and open sores in the Legs, Bones, and whole Body, together with all such distempers both inward and outward, by what name soever they may be called, as are occasioned by bloody *Mars*, the names whereof I shall forbear to recite, because these distempers and diseases, which are subject to

Mars and under his jurisdiction, are confessed by and known to experienced and skillful Physicians.

This Spirit of Iron being rightly discerned and known, has a secret affinity with the Spirit of Copper, that they may be so joined together, that there rises one only matter from them, of one and the same operation, form, substance, and being, which will cure and relieve the same distempers, and transmute the particularities of Metals with profit and honor. But Iron together with its virtue ought properly to be considered in the manner following, that it has a terrestrial body only in its corporeal form, which body may be used to a great many things, to alter the blood, to outward wounds, to a graduation of Silver, and inwardly to the constipation of the body; which yet is not always beneficial to use, neither in a man's body inwardly and outwardly, nor yet as concerning Metals. Because there is no great advantage to be made *per se* without the known right means, which do belong to Natures Secret Knowledge.

I must remember one thing more, that the lodestone and the true Iron are almost of one and the same use in bodily distempers, and are almost of one and the same Nature, even as it is according to a Divine, Spiritual, and Elemental sense, betwixt the Body, its Soul, and the Chaos, out of which the Soul and Spirit are gone, the Body is framed last of all out of that composition. What shall we do now, the ignorant and rude will not apprehend it, those as understand something will take notice of my writings, and those Nature surpassing wits will find fault. Here I do want some advice, for I would fain find out a way, that all these mighty wise men might continue my friends: Which I shall declare to you thus, that, because the argument itself does declare and pronounce the sentence and conclusion, there the resolution remains open, and cannot come further under any judgement of the mind, but must be declared, resolved, retained, and signified by itself.

Finally, observe this in this Chapter, that no House-keeping can be rightly and constantly performed betwixt a married couple, where one party will turn and drive the Chariot towards the East, the other towards the West. For they possibly cannot draw after this fashion, the said Chariot in an equal poise, whence there arises a great dissension and hindrance to obtain that which they imagined. But if such Married Persons do intend to govern their House well, they must be of one spirit, one opinion, one mind, and virtue to perform and act all, that is in their hearts and minds to work one with the other, if so be their love and faithfulness is perfect. For want of one of these parts, the three Principles are not rightly together. For Mercury is full of cowardice, and too little, as concerning constancy and fixedness. Sulphur is too little, it cannot heat the body of love, it being too much quenched. And the Salt has not its due, fit, and natural kind either, but is too hard, and too much, and therefore is the cause of a hard coagulation, is sharp and fretting, and does not show itself by fidelity and constancy. This is the course of this World, and the World is big of this vice, for there is but little constancy, small love, and less fidelity. This Philosophical example, I hope, will not be misconstrued, because *Syrach* extolls and dispraises the faithfulness and malice of a false woman, and both in a different manner. With this I bid *Mars* farewell, because no man knows to distinguish things of one and the same nature, much less such as are of a different, but he that has thus observed them, and has made a strict enquiring into their nature and properties, and by such accurate enquires has found them out and learned them. God the Father of Heaven, and the everlasting power, which yet was from the beginning, separate us in such a manner, that this terrestrial and corruptible body may attain unto, receive and comprehend the heavenly, spiritual, and incorruptible clarification.

If you cannot discern and know Iron by itself, give it a help meet for it. Judge then, and you will discover its power.

Chapter VI. Of the Spirit of Gold.

The brightness of Heaven has now commanded me to change my Pen, to discover a thing of Fortitude and of Constancy, because the Sun is a burning consuming fire, hot and dry, wherein there is a true power of all natural things, which power of the Sun works wisdom, riches, and health. My heart is seized upon by sorrow, and my spirit within itself becomes astonished, to manifest and to bring to light such things, as have not been discovered and commonly laid open before me, and to reveal that which has been buried in the depth with the very great secrecy. Notwithstanding, if I do go into myself, and examine my conscience, I could not find any other alteration to turn my mind, and to bring it upon some other design, which might obstruct my former resolution, but I shall speak with discretion, and write with prudence, that no evil may be occasioned by it, but rather some grateful good obtained, which I have delivered in the same manner, as other Philosophers have done before me.

Observe well, and having fixed your thoughts, put aside all these strange things, which are not serviceable to your speculation of Philosophy, but rather do destroy that advantage, which you have so earnestly sought for. Know then, that if you do earnestly long for, and heartily desire to get this golden Load-stone, your prayers first of all must be rightly made to God, in true knowledge, contrition, sorrow, and true humility, for to know and to learn the three different Worlds, wherein the immortal soul keeps its seat and residence, besides its first original, and is by Gods creation the first moving sensibility, or the first moving sensible soul, which of a supernatural being has wrought a natural life; and this soul, and this spirit, is the root, and the first fountain, and the first creature existing in the life of anything, and the *primum mobile*, which has been controverted so much by learned and very wise men. Furthermore, observe likewise the second Celestial World, and take very good notice of it, for therein do reign the Planets, and all heavenly Stars have their course, virtue, and power in this heaven, and do perform therein their service, for which God has placed them there, and do work in this their service by their Spirit, both Minerals and Metals.

Out of these two Worlds arises another different World, wherein is found and comprehended, what the other two Worlds have wrought and produced, out of the first supercelestial World is derived the fountain of life, and of the soul: From the second Celestial World, does spring the light of the Spirit; and from the third, the Elemental World, comes the invincible, heavenly, yet sensible fire, by which is digested and is ripened that which is comprehensible. These three matters and substances do generate and bring forth the form of Metals, amongst which Gold has the pre-eminency, because the Sidereal and Elemental Operation has mellowed and ripened the Mercury in this Metal the more substantially, to a sufficient and perfect maturity. And as the Seed of a Man does fall into the womb, and touches the *Menstruum*, which is its earth, but the Seed, which goes out of the Man into the Woman, is wrought in both the Stars and Elements, that it may be united and nourished by the Earth to a generation: So you are likewise to understand, that the soul of Metals, which is conceived by an unperceivable, invisible, incomprehensible, abstruse, and supernatural Celestial composition, as out of Water and Air, which are formed out of the Chaos, and then further digested and ripened by that heavenly Elemental light and fire of the Sun, whereby the Stars do move the Powers, when its heat in the inward parts of the Earth, as in the womb is perceived: For by the warming property of the Stars above, the Earth is unlocked and opened, that the infused spirit of the same may yield food and nourishment, and be enabled to generate something, as Metals, Herbs, Trees, and Beasts, where every one particularly brings with it its Seed for a further multiplication and augmentation. And as the conception of a Man is spiritual and heavenly, whose soul and spirit by nourishment of the Earth in the mother's womb, are formally brought up to a perfection: So likewise, it is to be observed and understood in every particular of Metals and Minerals. But this is the true Secret of Gold, *viz.* to instruct and teach you by an example and similitude, whereby the possibility of Nature, and its mystery is to be found in the manner following. It is probably true, that the heavenly light of the Sun is of a fiery property, and of a fiery being, which the most high God, as Creator of Heaven and Earth has granted to it, through a heavenly, constant, and fixed sulphureous spirit, for the preservation of its substance, form, and body; which creature by reason of its swift motion and course, through its swiftness is inflamed, and set on fire by the Air; which inflammation will never be extinguished, as long as the motion does last, and the whole created visible World does continue and endure, nor in the least diminished in its power: because there is no combustible matter extant, which might be given to it, whose consumption might cause the decay of that great light of Heaven. So is Gold by the Superior of its Essence thus digested and ripened, and is become of such a fixed invincible nature, that nothing at all can hurt it; because the upper fixed Stars have penetrated the lower, that the lower fixed Stars, by reason of the infusion and grant of the upper, need not to give place to their equal; because the lower has received and obtained such a constant fixedness from the upper. This now is very well to be noted and observed, as concerning the first matter of Gold.

I must allege yet another similitude according to the manner of Philosophers, of the great light of Heaven, and of that small Fire, which being terrestrial is here kindled every day, and is made to burn before our eyes. Because that great light has a magnetical likeness, and an attractive loving power with that small fire here upon earth, which yet is without form and impalpable, and found only spiritual, invisible, insensible, and incomprehensible. It is remarkable, as it is proved and demonstrated by experience, that that great light of Heaven has a great love for, and bears an affection and inclination to the little fire, which is terrestrial, by reason of the spiritual Air, whereby both are agitated, and preserved from their utter ruin and destruction. For do but consider, as soon as the Air, through great moisture or humidity, which it has attracted and taken in, conceives any corruption, that so through mists, and further coagulation and conjunction, clouds are generated., the beams of the Sun are hindered and obstructed, that the Sun cannot obtain its reflexion, nor have its due penetrating and searching power: So likewise this little terrestrial fire does never burn so clear in dark, cloudy, and rainy weather, neither does it show itself with that gladness in its operation, as when the air is fair, pure, clear, and unmixed, and heavenly. The cause is this, for though the obstacle of the moist air, the love is hindered and obstructed, that the attractive power growing sad, cannot exercise its perfect love and operation, as it ought to do, for the contrary element, the aquosity, causes this obstruction. As now the Sun, that heavenly great light, has a special communion and love with the small terrestrial fire, to attract after a magnetic manner: So likewise has the Sun and Gold a special correspondency, and a peculiar attractive power and love together, because the Sun has wrought the Gold through the three principles, have their Lodestone, which is nearest of all related to the Sun, and has attained to the highest degree, so that the three Principles are found most mighty and powerful in the same. Next to it is Gold in its corporeal form, because it is framed out of the three Principles, but has its original and beginning from the heavenly and golden Loadstones. This is the greatest wisdom beyond all wisdom nay, a wisdom beyond. all human reason and understanding. For by this wisdom is first of all apprehended the Creation, the heavenly Essence, the operation of the Firmament, the spiritual imagination, and corporeal being, and does comprehend all the qualities and properties, and whatsoever does maintain and preserve a man. In this golden Loadstone is and lies buried the dissolution and opening of all the Minerals and Metals, their government, as also their matter of the first generation, and their power, as touching health. Moreover, the coagulation and fixation of Metals, together with the operation to cure all diseases. Take a special care of this Key, for it is heavenly, sidereal, and elemental, out of which the terrestrial is generated. It is Supernatural and Natural together, and is born out of the Spirit of Mercury, heavenly; out of the Spirit of Sulphur, spiritual; but out of the Spirit of Salt, corporeal. This is the whole way, and the whole substance, the beginning, and the end. For the Spirit and Body are so knit together in one by the Soul, that they can never be separated, but do generate a most perfect and fixed body, which can receive no hurt.

Out of this spiritual Essence, and out of this spiritual Matter, out of which the Gold first of all is made corporeal into one body, is the Potable Gold more substantially to be made, then out of Gold itself, which must be made spiritual, before the Potable Gold can be prepared out of it. This Spirit cures likewise the Leprosie, the French-pox, as being a superfixed Mercurial Essence, dries up and consumes the Dropsie, and all running open sores, which have afflicted a long time, comforts the Heart and Brain, strengthens the memory, breeds good blood, and causes gladness and desire, and natural enticement to Carnal Copulation. If you mix the Quintessence of Pearls with the tincture of Corals, and do administer and join them with it, the same quantity of this Spiritual Essence of Gold, the quantity of two grains, you nay assure yourself, and boldly rely upon it, that no natural distemper can trouble you, or do you any hurt to endanger your health. Because the nature of the Spirit of Gold is to change and alter all infirmities, to take them away, and to cure them, that so the body may become perfect without any distemper. The quintessence of Pearls does comfort the heart, strengthening the memory and senses. The Tincture of Corals dispels all poison together with the evil Spirits, which do abhor that which is good.

Thus, can the soul of Gold reduced into water, the spiritual Essence of Pearls, and the Sulphur of Corals united in one do such things, which to Nature seem otherwise incredible. But because experience does confirm the infallible truth, it is then deservedly a Cordial in this mortal life, and is justly, by reason of its wonderful effects, preferred before all the Cordials, by what names soever they may be called. I am a Spiritual man, subject to the Ecclesiastical State, and engaged by a Spiritual and Divine Oath to the Order of the Benedicts, whereby through my devout prayers I do receive great and precious promises of the Word of God, to the comfort of my Soul: But in corporeal afflictions of my infirmity, as well as of my Brethren, I have not found nor used a greater cordial by God's blessing, than this composition of the three things aforesaid. God grant, and increase this power and. virtue to the end of this temporal world, which men exchange for death.

O thou golden power of that Soul of thine, O thou golden understanding of thy Spirit! O thou golden operation of thy body! God the Creator preserve thee, and. grant to all terrestrial creatures which love and honor him, the true knowledge of all his gifts, that his will may be done in Heaven, and on Earth. And thus, much may suffice for the present for the discovery of the Spirit of Gold, till the coming again of *Elias*.

To this I will add yet this short Process.

Take Spirit of Salt, and with it extract the Sulphur of Gold. Separate the Oil of Salt from it, and rectify the Sulphur of Gold with Spirit of Wine, that it may become pleasant, without any corrosiveness. Then take the true Oil of Vitriol, made out of the Vitriol of Verdigreese, dissolve it in Iron: Make again a Vitriol out of it, and dissolve it again into an Oil or Spirit, which rectify likewise as before, with Spirit of Wine. Put them together, and draw off the Spirit of Wine from thence. Dissolve the natter, which remains dry behind, in the Spirit of Mercury, in a due proportion or weight. Circulate and coagulate it. When it becomes constant and fixed without rising any more, you have then, if you ferment it with prepared Gold, a Medicine to tinge both Men and Metals.

Chapter VII. Of the Tincture of Silver.

The Tincture and Spirit of Silver is of a Sky-color, otherwise it is a waterish Spirit, cold and moist, and not so hot in its degree, as the Spirit which is found in Gold, Iron, or Copper, therefore is Silver more phlegmatic than fiery, although it has been reduced by fire out of its waterish substance unto coagulation. In what manner Metals do obtain their tinging Spirits and coagulation; in the same manner have the Stones likewise received their hardness, fixedness, and Tincture, as by one and the same influence. In a Diamond is found a fixed and coagulated Mercury, therefore this Stone is harder and more fixed than other Stones, and is not to be broken as they are. In a Ruby is found the Tincture of Iron, or the Sulphur of Iron. In an Emerald, the Tincture of Copper, in a Granate the Soul of Lead. In Pewter, the Tincture which is found in the stone called Topaz. Crystal is attributed to common Mercury. And in a Sapphire, is found the Tincture and Sulphur of Silver; yet everything in particular, according to its nature and kind; and in Metals likewise according to their form and kind. And when the blue color is separated and taken away from the Sapphire, then is its garment gone, and its body is white like a

Diamond. Thus, when Gold has lost its Soul, it yields then a white body, and a fixed white body of Gold, which is called *Luna fixa* by the searching students and novices of this Art.

What has been said concerning the stone called Sapphire, for your instruction, you may apply to the better knowledge of the nature of Metals. For this blue Spirit is the Sulphur and Soul, out of which Silver has its life, as well in the Earth, as above the Earth, by Art, and the white Tincture of Silver, upon white always, in a magnetic form of that one thing and Creature, wherein the *Primum Ens Auri* likewise is found.

You most Eminent Orators, where is here your Rhetoric to declare this Mystery? And you reputed Enquirers into Natures Secrets, what is become of your writings and knowledge? You Physicians likewise, where is your opinion and judgement gone? Perhaps beyond Seas, to fetch something from far remote places, wherewith you may cure the Dropsie, and other lunary distempers. You will say, my speech is too dark for you: If so, kindle then that Terrestrial fire, and search and be not ashamed, to make friendship with Vulcan, let no painstaking discourage you, and you will find by the permission of the Eternal God, that the Spirit of Silver all alone contains that, which will perfectly cure and dispel the Dropsie: Even as the Spirit of Gold, and of Mercury, can radically cure the Consumption, so that even the center itself of the said distemper may not be found.

But that Silver is not so provided in its degree with a hot substance and quality, in the veins of the Earth, but is subjected to a waterish kind; this fault is to be laid upon the great light of heaven, which by reason of its waterish influence has planted this quality into the second Creature and into the second Planet of the Earth, as unto Silver. And though Silver does carry with it a fixed Mercury, or fixed Quicksilver, which is born in it: Nevertheless, it wants the hot fixed Sulphur, which might have exactly dried up and consumed the phlegm, which is the cause it has not obtained a compact body, except it be done afterwards by Art of the lesser World. And because the body is not solid and compact, by reason of its waterish substance, hence are its pores not well stopped up, nor consolidated, that it might have a due ponderosity, and endure a fight with its enemies. Which virtues ought altogether to be found in Gold, if so be it must conquer all its foes, and endure all the trials without fault. Al things are hard and difficult in the beginning, but when they are brought to a period, they are easy to be understood and comprehended.

If you do rightly observe and learn to know the Spirit and Soul of Silver, you will easily apprehend the main work, how they must yield the end of their usefulness. Therefore I will propose unto you an example, and instruct you by a Country rule, what you are to apprehend and to consider from children's play, to things of great concernment, that you nay advantageously enquire into, and meditate upon them, *viz.* a common Countryman sows Linseed upon a well dunged, and well tilled ground, which Linseed comes forth out of the earth after its putrefaction, by the operation and furtherance of the Elements, and lays before our eyes a matter of Flax together with its Seed, which it brings increased with it, which Flax is plucked off and separated from its Seed. This Flax cannot be used, nor prepared with any profit for any work, except it be first of all putrefied through Water, by which putrefaction the body unlocks itself, and gains an ingress of its usefulness. After this putrefaction and unlocking is performed, the flax is dried again by the Air and the Sun, and through the coagulation reduced and brought again to a formal being, that it may further serve for work. This prepared Flax is washed, beaten, broken, and swinged, last of all heckled, that the *purum ab impuro*, the clean from the unclean, the gross from that which is subtle may be separated, which otherwise is impossible to be done and performed, except the aforesaid preparation go before. Afterwards they Spin this Flax, and it becomes Yarn. This Yarn is boiled in Water upon the fire, or laid into Lees and placed by the fire, whereby a new purification happens, that so the impurity and superfluous uncleanness may be further severed and separated. After the Yarn is rightly washed and cleansed, it is dried, and carried to a Weaver, and Cloth made of it; which Cloth by often wetting with water is clarified and whitened, cut in pieces by Taylors, and other people, and used according to every housekeeper's necessity. And after such Linen cloth is torn and worn to pieces, and as it were reduced to nothing, then are the old rags picked, and gathered, and carried to a paper-mill, where the Master makes Paper out of them, which may likewise be used for several things. This Paper being laid upon some Metal or Glass, and then lighted and burnt, the Vegetable Mercury goes then out of the Paper into the Air, and flies away. The Salt remains in the Ashes, and the burning Sulphur, that which is not so quickly consumed in the burning, is resolved into an Oil, which is an excellent Medicine for dark and bad Eyes. This resolved Oil has in it an extraordinary fatness, which the matter of the Paper has carried with it from the beginning of its first Seed of Flax: And thus the last matter of Flax, as Paper, is again resolved into the first matter, as into the pingued[25] sulphureous oleosity of the Linseed, together with a separation of its Mercury and Salt; that so out of the last, the first as the foundation, is made manifest, and out of the first the virtue and operation is known and learned.

Though this Argument be rude and gross, yet will you learn by it things secret and subtle. For that which is subtle, must be infused under gross examples into ignorant people: For they are to learn, to put off that which is gross, and to take in that which is subtle. Thus, you are likewise to understand, that the first matter of Metals must be observed, studied, and found out, through the discovery of their last

---

[25] Exact same spelling in the 1671 edition. -pnw

matter, which last matter, as there are the absolute and perfect Metals, must be divided and separated, that it may appear altogether naked to a man's Eyes, and then there may be learned & known by such a division, what the first matter has been in the beginning, out of which the last is made. Thus, much concerning Silver. I had several things yet to mention, but I will leave them for another time: most heartily beseeching you, and exhorting you upon your conscience, that you will observe all these things which I have discovered unto you, and all these letters which are comprehended betwixt *Alpha* and *Omega*, and carefully keep all my sayings and writings, that you may not crave pardon for your sins, and suffer everlasting vengeance to eternity. Lastly, I do impart this yet unto you. Take the Sky-colored Sulphur of Silver, which has been extracted out of Silver, and rectified by Spirit of Wine. Dissolve it according to its weight in the white spirit of Vitriol, and in the sweet-scented spirit of Mercury, and coagulate them together through a fixation of fire, and you will get the possession of the white Tincture, and its Medicine. But if you know the *Primum Mobile*, it is then needless, because you may bring the work to perfection out of one.

Chapter VIII. Of the Soul or Tincture of Pewter.

The benign *Jupiter* is almost of a middle nature amongst all the Metals. He is neither too hot, nor yet too cold, not too warm, nor too moist. He has not too much Mercury, nor yet of Salt, and of Sulphur there is least of all in him. Pewter is found white in its color, yet of these three principles one does exceed the other, as it has been clearly discovered in its division, accord to the true enquiry into natures secrets. Out of this composition and mixture of the three principles, is generated, and wrought, and coagulated into a Metal, and brought to a maturity of perfection, *Jupiter* a God of peace, a bountiful Governor, and a Lord and Prince of the middle region, as concerning his estate, essence, profession, virtue, form and substance; for he keeps the mean, and there can hardly happen any distemper, where *Jupiter* may do any extraordinary hurt, if his Medicine be soberly used in not too great a quantity. It is likewise reputed needless, where his Medicine is not required, to administer it, not being called for, to strange things; but is justly reserved for these where the body and distemper have a similitude of likeness with the upper stars, and their help in virtue, power, and operation, and so agree together in their conjunction, that there may not be found any contrariety, neither in the operation, nor in the operating nature. The Spirit of *Jupiter* is found such; that it may in no wise be spared in the generation of Metals, as likewise no Spirit of any Metals may be put by, because they must need concord and agree from the lowest to the highest degree, if there shall follow a perfect metal in the Earth, and likewise in the little World by transmutation and augmentation. This is now to be understood thus, that all degrees, from the lowest Metal to the Highest in perfection, must be gone through, as likewise the Metals must perform their course from the very Lead to Gold, by reason of the fixedness of the tincture and body, notwithstanding that *Saturn* keeps the chiefest place in the highest region, wherein the Astra do reign, and the Stars do perform their motion. The generation of Pewter in and above the Earth, is brought to light after the following manner. As a Man, and other Beasts are first brought up and fed with the milk of their Mother, and there is no meat to be found upon the face of the Earth, more convenient and fit for the bringing up of Men than milk; for their melioration is for the most part by an animal Sulphur, which yields the nourishment. Thus, likewise is Pewter fed and brought up by its metalline Sulphur, which is most agreeable to it, and sucks in more warmth and heat than Saturn, and therefore is *Jupiter* the more digested and roasted, whereby also his body becomes more constant and fixed in the degree of Salt.

Jupiter orders that his Lordship and Dominions may be well governed, and Justice be rightly administered to everyone in his Court of Judicature. The Spirit of *Jupiter* does protect and preserve from all distempers and diseases incident and hurtful to the liver. Its Spirit is naturally, as for its taste, like unto honey. Its Mercury being made volatile, does get a venomous quality, for it purges vehemently, and penetrates with violence. Therefore, it is not always good, that its unlocked Quick-silver should be thus simply used by itself; but if a correction goes before, it may be very well used with exceeding great usefulness in those distempers and diseases, which are immediately subject to his influence; that is to say, when you have taken away from *Cedekiel* its venomous volatility, and it is placed into a better and more fixed estate, which does resist poison.

This description will transcend the capacity of a common physician, because this Art and Science does not consist in bare words, but comes by experience: in words has a common physician his end and foundation; but the preparation of our Medicine does begin with words, but its foundation is in a certain trial to be made by experience, which foundation is laid upon a hard rock by manual operation, but the other upon a shaking Reed, and mere Sand; therefore that which is firm and unmovable, made by the hands of Nature, is justly preferred before bare words, which do flow from an inconstant fantastic speculation, because the workmanship commends always its Master.

I do not speak now after my poetical manner, nor do I write in that style, which I have used in discovery of that miraculous nativity of the seven planets in my *Occulta Philosophia*; neither do I make use of a magical or cabalistical manner; much less do I observe the Method of these, which have industriously studied and learned the secret, hidden, and supernatural Arts, as there are *Hydromantia, Aeromantia,*

Geomantia, Pyromantia, Nigromantia, and the like: But my purpose and intention for the present is to discover the secrets of Nature, that the Lovers of Art, and children of this searching and desired wisdom, may, by the blessing, mercy, and permission of God, apprehend, observe, and study them, and after diligent observation may learn and keep some useful things, as well concerning the generation of Metals in two parts, in the greater and lesser World, as also the true Medicine, which these Metallic and Mineral forms do contain in their inward parts, which are apprehended and demonstrated by dividing, that their first beginning is notoriously manifested in three several things. Then is Nature uncovered, and the secret parts are laid open by putting off the temporal cloths, and all its secret virtue, power, and operation discovered for the good and health of Mankind. My persecutors and unskillful physicians will say, you speak much of Geese, and I do not know the Ducks yet; who knows whether everything be true, which you have set down for truth in your writings. I will stick to the things I have experience of, and which are practiced by my fellow physicians; thus shall I continue undeceived, and I do assuredly know, that I need not take any pains for to learn novelties. He that is of such a resolution, will certainly abide with his Ducks, and never deserve roasted Geese, to learn the Secrets of Nature.

But I do sincerely confess, and call the most High Trinity to witness, under the loss of the noblest spiritual jewel, that whatsoever I have written, and shall yet write here, is truth, and will be found to be nothing else but the truth. But that every understanding, and every common man, but especially those that are haters and persecutors of these secrets, do not understand my writings so well and so clearly, this, I say, I cannot help. Pray you to God for his grace, and your persecutors ask him forgiveness, work cheerfully, and use your reason when you read, and there will be no secret too abstruse for you, but you will by enquiry find it out. Moreover, I do exhort you, that whosoever finds out this gift of God, may without intermission, day and night, offer most hearty thanks with all humility, and reverence, and due obedience to God the Creator chiefly; because no Creature is capable to render sufficient thanks, which might equalize this benefit: Therefore, diligence is to be known by a true and real industry, per one's power. I have done my due, for which I intend to answer to God and the World, for what these eyes of mine have seen, my hands have felt, and myself with reason infallibly comprehended, shall no body take away from me in this life, death excepted, which does separate all things.

I have not indeed been forced to write these things, but whatsoever I have done, I have neither done out of rashness, nor yet out of a desire to get to myself a temporal lasting name; but I have been put upon it by the command of Christ the Lord, that his Majesty and Mercy, in Eternal and Temporal things, may not be hid from any man, but may be manifested to the praise, honor, and glory of his holy and everlasting Name, that it may be, by reason of its omnipotency, confirmed by the performing of great miracles, magnified, acknowledged, and in its majesty exalted.

The second thing that moved me to write this, was Christian charity to my neighbor, to do him as much good as to myself, and thereby to heap fiery coals upon the heads of mine enemies.

And lastly, that all my adversaries likewise might acknowledge, what errors or by paths others had been in, in comparison to me, and who had most faithfully discovered Nature's *Arcana*, whether I am to be condemned, or they be justified. And then last of all, that the highest mystery might not be altogether choked in darkness, nor overwhelmed by the swelling waters, but being freed from the miry and filthy ditches of a simple and ignorant crew, might get many witnesses, by spreading abroad of a true, certain, and due acknowledgement, to follow my footsteps in discovering the truth.

They attribute to me in my Lordship, of the 12 Signs of the Stars of Heaven, the Archer and Fishes, out of these I am born, because I was in an aquosity before I began to live; but the Archer has fixed his Arrow into my heart, that I having lost this aquosity of mine, was made worthy of the dry earth; And though the earth was brought into a soft substance by the Water, yet are you to know, that the water was consumed through the dry and warm air, and so all the soft matter of the Earth being vanished, I was made worthy to receive that hardness by the aforesaid exsiccation.

By this, as well the Scholar as the wise man, are to take good notice and diligently to observe, that *Jupiter*, as well as other the chiefest planets, is subject to the four Elements, which Elements have received their Centrum from above, and are born as others are.

To conclude all, I will let you know, that if you do extract out of the benign *Jupiter* his Salt and Sulphur, and make Saturn flow very well together with them, Saturn does get a fixed body, is purified, and becomes clear by them, and is a total change, and real transmutation of Lead into good Pewter, as you will find it upon a most accurate trial. And although this may seem to you not to be true, yet are you to understand, that by reason that the Salt of *Jupiter* is made more corporeal only by its Sulphur, it likewise has received an efficacy and a power to penetrate SATURN, as the vilest and most volatile metal and to bring it to its own substance by making it better, as you will really find it to be so.

Chapter IX. Of the Spirit of Saturn or Tincture of Lead.

Saturn to generate his metal, which is Lead, is placed in the upper Heaven above all Stars. But in the lower parts of the Earth he does keep the lowest degree. As the uppermost light of *Saturn* is mounted to the highest altitude of all the lights of Heaven; so likewise, in imitation of the same, has Nature given leave and permitted, that his children of the lower Region have retired themselves by *Vulcan* to those of their quality, according as *Saturn* has been moved. For the upper light is the cause of it, and has generated an unfixed body of Lead, through which go and are drawn open pores, that the air can have its passage through this Saturnine body, and bear it up. But the fire easily works upon and consumes it, because the body is not solid and compact by reason of its unfixedness. This is well to be observed by a serious enquirer into all things; because there is a vast difference betwixt fixed and unfixed bodies, and then the causes of this fixedness and unfixedness: And though *Saturn* is of a singular ponderosity before other Metals, yet will you observe, that when they are poured out together, after their conjunction in the melting of them, the other Metals will always fall to the bottom, as likewise it happens with other metals, by pouring them through Antimony. Whereby it does appear, that other metals have a more solid and compact body, then *Saturn* can raise; because it must give place to other Metals, make room for them, and yield the victory: for it vanishes away, and is consumed together with those unconstant and unfixed Metals. For there are the three grossest qualities of the three principles in *Saturn*, and by reason that its Salt is altogether fluid in comparison to other Metals and Planets, therefore is likewise its body more fluid, inconstant, unfixed, and more volatile, than any Metallic body.

How *Saturn* does proceed towards his regeneration, you are to know, that as common Water through natural cold, by the alteration of the upper Heaven is congealed, so that it becomes a coagulated Ice: So likewise, it is demonstrated, that Lead is coagulated and made corporeal by reason of the great cold, which is found in its Salt before any other Salt. The congealed Ice is resolved through warmth, and so is the coagulated Lead made fluid by fire. It has most Mercury in it, yet inconstant and volatile: But less of Sulphur; and therefore, according to the small quantity of the same, its cold body cannot be heated; and least of all of Salt, but fluid: Otherwise the iron would be more liquid and malleable than *Saturn*, if the Salt alone could impart both the malleableness and fluidity, because Iron does carry with it more Salt than any other Metal. And being there is a difference to be found in these things, you must carefully observe how Metals are to be distinguished. All the Philosophers indeed, besides myself, have written, that the Salt causes the coagulation, and the body of every Metal: And this is true, but I shall let you see by an example, how this is to be understood. *Alumen plumosum* is reputed and probably accounted to be a mere Salt, and herein may be compared to Iron, which Salt is the aforesaid *Alumen plumosum* is nevertheless found to be as a matter, and not liquid like Iron. On the contrary, Vitriol does show itself like Salt in a small quantity, yet liquid and open; and therefore, its Salt cannot cause so hard a coagulation in its appropriated Metal, as that other Salt does. Although all the Salts of Metals are grown out of one root, and one seed, yet is there a difference of their three principles to be observed: As one herb differs from the other, and likewise in Men and other Beasts a difference is found as concerning the original of their qualities, and their three principles, where one Herb has something more of this, another Herb more of that kind; which is likewise to be understood of Men and Beasts. The soul of Lead is of a sweet quality, as also the Soul of *Jupiter*, and yet sweeter, so that as for sweetness there is hardly anything comparable to it, being first highly purified by separation, that the pure being very well severed from the impure, there may follow a complete perfection in the operation. Otherwise the spirit of Lead is naturally cold and dry, therefore I do advise both Men and Women not to make too much use of it, for it over-cools Human Nature, that their seed cannot perfect or perform its natural operation; nor is it good for the spleen and bladder, it does attract the phlegmatic quality, which breeds melancholy in men. For *Saturn* is a Governor, and such a melancholy one, whereby a Man is upheld and strengthened in his melancholy: Therefore, if its spirit be used, one melancholy spirit does attract the other, whereby a man's body is freed and released from its infused melancholy. Outwardly is the soul of *Saturn* very wholesome in all sores and wounds, whether they be old or green, whether they happen by thrusting or cutting, or naturally by means incident, so that hardly any other metal will do the like. It is a cooling thing in all hot and swelled Members; but to eat away, and to lay a foundation for healing in all corrupt and putrefied sores, which have their issuing forth from within, there the noble *Venus* has the pre-eminency; because Copper is hot in its Essence to exsiccate and dry up, but Lead on the contrary is found to be cold in its Essence.

That heavenly Light of the Sun is much hotter than the Light of the Moon, because the Moon is much lesser than the Sun, which does comprehend the eighth part of magnitude in the circle of measuring and dividing. And if the Moon should exceed the Sun in this magnitude of the eighth part, as the Sun does exceed the Moon; then all the fruit, and whatsoever grows upon the earth would be spoiled, and there would be a continual Winter; and no Summer would be found. But the eternal Creator has herein wisely prescribed a certain order and law to his Creatures, that the Sun should give light by day, and the Moon by night, and thus be serviceable to all Creatures.

Those Children which are addicted to the influence of *Saturn* are melancholy, surly, always murmuring, like old covetous misers, which do no good to their own bodies, and are never satisfied; they use their bodies to hard labor, vex and fret themselves with troublesome thoughts, and are very seldom so cheerful as to recreate themselves with other people, neither do they care much for natural love of handsome and beautiful women.

To sum up all, I do tell you, that *Saturn* is generated out of little Sulphur, little Salt, and much immature and gross Mercury; which Mercury is to be accounted like scum or froth, which swims upon the water, in comparison to that Mercury which is found in Gold, being of a much hotter degree. Hence it is, that the Mercury of *Saturn* has not so fresh and so running a life, as that which is made out of Gold, because more heat is found in this, to which the running life owes its original. Therefore, it is likewise to be observed in the inferior world of the little *Vulcan*, in the augmentation and transmutation of Metals, what description I have given you of these three principles of *Saturn*, concerning their original, quality, and complexion. And everyone is to know, that no transmutation of any Metal can arise from *Saturn*, by reason of its great cold, except the coagulation of Mercury: because the cold Sulphur of Lead can quench and take away the hot running spirit of Quick-silver, if the Process be rightly performed; therefore, it is rightly to be observed, that the Method be so kept, that the Theory may agree with the practice, and concur in a certain measure and concord. Wherefore you must not altogether reject *Saturn*, nor vilify and disparage it; for its nature and virtue is not as yet known but to few. For the Stone of the Philosophers has the first beginning of its heavenly resplendent Tincture only from the Metal, and by infusion of this Planet, is the key of fixedness delivered to it through putrefaction; because that out of the yellow there cannot come any red thing, except there be first made out of the beginning of the black a white one.

There are yet many things to be treated of, as of many miracles of natural and supernatural things; but because some other business has hindered me from making a more ample and fuller relation of them, I shall here conclude this treatise of mine; and the rest concerning the hidden secrets of Minerals, you shall have in other writings, *viz.* in a peculiar book of Antimony, Vitriol, Sulphur, Loadstone, and which more especially before others are endowed, and do adhere to them, from which Gold and Silver derive their original, middle, and end; together with the true transmutation in particular, which their virtues, powers, and efficacies they have received out of one thing, wherein these, together with the rest of Metals invisibly to be generated, are hidden, which matter is clear and plain enough before any man's eyes; but because its virtue, power, and efficacy, lies buried very deep, and is unknown to the greater part, therefore is this matter likewise accounted and esteemed, abject, vile, and unfit, for want of true knowledge of it, till the Disciples of Christ travelling toward *Emahus*, have their eyes opened, that they discern by the Breaking of Bread, what wonderful miracles the rich Creator has planted and put into a contemptable Creature: Its Name is *Hermes*, and in its Arms there is a flying Serpent, which is its Wife, and is called *Aphrodita*, that can search all men's hearts; and yet it is all one, and one only thing, and one only Being, which is common everywhere, and known in all places; everyone does touch and handle it, and does use it to things base and vile. Man does highly esteem that which is of small value, and rejects that which is high. It is nothing else but Fire and Water, out of which the Earth with assistance of the Air is generated, and by which it is as yet preserved. I do give most hearty thanks to the most High for his gifts. And thus, having discovered enough, according to the resolution I had taken in the Treatise of mine, I take my leave.

All will be found in the Separation.

FINIS.

A Word from the Publisher

Thank you for purchasing this volume from The R.A.M.S. Library of Alchemy. During his lifetime, Hans Nintzel was dedicated to the identification, acquisition, study, retyping and, when necessary, translation of what he considered to be the most important known works on Alchemy. Hans was assisted by his sparse network of fellow Alchemists, all members of the Restorers of Alchemical Manuscripts Society (R.A.M.S.). I was an active member of R.A.M.S.

My goal is to publish all the works originally made available through R.A.M.S. as photocopies.

The works from the original R.A.M.S. Library are republished by R.A.M.S. Publishing Company in the collection, "The R.A.M.S. Library of Alchemy," with permission of the Estate of Hans W. Nintzel.

If you have a work on Alchemy that you believe should be a part of the R.A.M.S. Library, please contact me through R.A.M.S. Publishing Company.

Philip N. Wheeler

The R.A.M.S. Library of Alchemy

The study and practice of Alchemy was extremely important to Hans W. Nintzel. He assembled this Library over a period spanning more than three decades, guided by his teacher Frater Albertus. The R.A.M.S. Library of Alchemy includes the most valuable Alchemical texts that Hans painstakingly located, acquired, retyped, and translated during his lifetime, with help from other R.A.M.S. members.

The following is a list of the volumes that are currently available or in progress. Volumes that contain works from multiple authors may have only the principle author or editor listed.

Volume	Title	Author or Editor
1	Twelve Keys of Basilius Valentinus	Basilius Valentinus
2	Triumphal Chariot of Antimony	Basilius Valentinus
3	His Secret Book	Artephius
4	The Golden Work	Hermes Trismegistus
5	Three Works of Ripley	George Ripley
6	Four Works of Paracelsus	Paracelsus
7	Bacstrom's Notebooks, Part 1	Sigismund Bacstrom
8	Bacstrom's Notebooks, Part 2	Sigismund Bacstrom
9	Summa Perfectionis	Geber (Abu Musa Jabir ibn Hayyan)
10	The Five Centuries	Rudolph Glauber
11	The Greater and Lesser Edifyer	Johann Grashoff
12	Chemical Secrets and Experiments	Sir Kenelm Digby
13	The Turba Philosophorum	Arisleus
14	Das Aceton	Christian Becker
15	The Art of Distillation	John French
16	Non-Violent Destruction of the Atom	Nintzel & Wheeler
17	Philosophical Furnaces	Rudolph Glauber
18	The Last Will and Testament	Basilius Valentinus
19	TBD	
20	TBD	
21	Alchemical Symbols, Third Edition	Hans W. Nintzel and Philip N. Wheeler
22	The Book of Formulas	John Hazelrigg
23	18 Short Tracts	Hans W. Nintzel
24	Bacstrom's Notebooks, Part 3	Sigismund Bacstrom
25	A Discourse on Fire and Salt	Blaise Vignere
26	The Mineral Work	Johan Hollandus
27	The Vegetable Work	Johan Hollandus
28	Lamspring's Process	Lamspring

29	The Book of Abraham the Jew	Abraham Eleazar
30	Five Short Works of Glauber	Johann Glauber
31	The Metamorphosis of the Planets	Johannes Monte-Snyder
32	Four Works of Roger Bacon	Roger Bacon
33	The Golden Chain of Homer	Homerus, Kirchweger, Nintzel, Wheeler
34	Alchemy Rediscovered and Restored	Archibald Cochren
35	Aurifontina Chymica	John Houpreght
36	The Golden Fleece	Salomon Trismosin
37	The Transmutation of Base Metals into Gold and Silver	David Beuther
38	Sanguis Naturae	Christopher Grummet
39	A Revelation of thye Secret Spirit	Giovanni Lambi
40	The Holy Guide, Part 1	John Heydon
41	The Holy Guide, Part 2	John Heydon
42	Secreta Alchymiae	Kalid Persica
43	The Golden Treatise of Hermes	Hermes Trismegistus
44	Potpourri of Alchemy, Part 1	Hans W. Nintzel
44	Potpourri of Alchemy, Part 2	Hans W. Nintzel
46	TBD	
47	Selected Chemical Universal and Particular Processes	Alexius von Ruesenstein

For the latest updates, please visit:

https://ramsalchemy.jimdo.com

www.ingramcontent.com/pod-product-compliance
Lightning Source LLC
Chambersburg PA
CBHW080234180526
45167CB00006B/2270

* 9 7 8 1 5 4 3 1 5 1 6 1 9 *